HUMAN POPULATION GENETICS

Foundations of Human Biology

Series Editors:
Matt Cartmill
Kaye Brown
Boston University

The Growth of Humanity by Barry Bogin

Fundamentals of Forensic Anthropology by Linda L. Klepinger

The Human Lineage by Matt Cartmill and Fred H. Smith

HUMAN POPULATION GENETICS

JOHN H. RELETHFORD

Department of Anthropology

State University of New York College at Oneonta

WILEY-BLACKWELL

A JOHN WILEY & SONS, INC., PUBLICATION

Published by John Wiley & Sons, Inc., Hoboken, New Jersey
Published simultaneously in Canada

Wiley-Blackwell is an imprint of John Wiley & Sons, formed by the merger of Wiley's global Scientific, Technical, and Medical business with Blackwell Publishing.

For general information on our other products and services or for technical support, please contact our Customer Care Department within the United States at 877-762-2974, outside the United States at 317-572-3993 or fax 317- 572-4002.

Wiley also publishes its books in a variety of electronic formats. Some content that appears in print may not be available in electronic formats. For more information about Wiley products, visit our web site at www.wiley.com.

Library of Congress Cataloging-in-Publication Data:

Relethford, John.
    Human population genetics / John H. Relethford.
        p. cm.
    Includes index.
    Summary: ''*Human Population Genetics* will provide an introduction to mathematical population genetics, along with relevant examples from human (and some non-human primate) populations, and will also present concepts and methods of population genetics that are specific to the study of human populations. The purpose of this book is to provide a basic background text for advanced undergraduate and graduate students interesting in the mechanisms of human microevolution''—Provided by publisher.
    ISBN 978-0-470-46467-0 (pbk.)
  1. Human population genetics. I. Title.
    GN289.R45 2012
    599.93'5—dc23

2011028962

Printed in the United States of America

10 9 8 7 6 5 4 3 2 1

CONTENTS

FOREWORD

If, like us, you find yourself hard-pressed to follow the fast-paced scrimmages of anthropological genetics from the sidelines, this is the book you have been waiting for. John Relethford, one of the world's leading contributors to these debates, has written it to engage all of us in this important and rapidly evolving area of scientific inquiry. In *Human Population Genetics*, he leads us through classic studies and current debates in an easy, clear, informal style that draws us in and involves us in the action and arguments. Relethford's passion for understanding the genetics of human populations, and his low-stress approach to what can be a difficult and esoteric topic, kindle a like passion in the reader and make this book that rare thing among textbooks—a source of excitement and inspiration.

Population genetics and statistical theory were born as conjoined twins in the monumental work of R. A. Fisher in the 1920s, which transformed evolutionary biology into a full-fledged science capable of making and testing predictions with numbers in them. But many people who are eager to learn about human biology and evolution are turned off by the statistical foundations of evolutionary theory. Almost everyone who teaches the fundamentals of our science has learned to dread the dazed expressions that come over students' faces the moment the Hardy–Weinberg equation hits the screen. Relethford shows us, and them, how to get around this stumbling block. Drawing the reader effortlessly in through plain and simple examples beautifully chosen to clarify the mathematics of probability, Relethford recruits his mastery of the subject and his skill as a teacher and writer to present the math in a user-friendly way that displaces the hard work of deriving formulas into adjacent appendices. His readers first master the essentials and later reward themselves by seeing the mathematics underlying the simple models they have just grasped. This process of orderly presentation leaves readers self-confident and ready to take on ever more complex material.

Throughout this book, Relethford systematically preaches and teaches a scientific approach to knowledge (*"Much of science consists of developing a simple model, testing its fit in the real world, and then explaining why and how it fits and does not fit"*) in a way that always solicits involvement by the reader (*"To see this, let us try an example"*). In every topic he presents, he returns to the readers' point of view (*"What effect do you think selection has had on the allele frequencies?"*) and includes them in the developing narrative. His readers will learn the concepts that are crucial to all fields of population biology by studying examples of special relevance to biological anthropology—how familiarity with genetic evidence can inform us of our history (see the rich discussion on tracking the appearance of the *CCR5-Δ32* allele and subsequent resistance to the AIDS virus), how adaptation has taken many different paths in human history (see the discussion on different high-altitude adaptations in Tibetan and Andean people), and how cultural behavior impacts genetic processes (see the discussion on agriculture and hemoglobin S). "Instead of cultural evolution negating genetic evolution," he writes, "we are finding evidence of how cultural change has accelerated genetic evolution."

That sentence, and the evidence behind it, would by itself make *Human Population Genetics* worth having on your bookshelf. Every chapter of the book sparkles with conclusions that are just as simple, straightforward, and far-reaching. All of its readers can rely on John Relethford to lead them into some of the most important and exciting scientific conversations of our day. If you are a student of biological anthropology at any level, or a scientist or educator who teaches these subjects, you will find his new book an invaluable source of novel insights and fresh illumination of key ideas. We are proud and delighted to see *Human Population Genetics* added to the Wiley–Blackwell series of textbooks on the foundations of human biology.

KAYE BROWN
MATT CARTMILL

PREFACE

WHAT ARE WE DOING HERE?

This book is about the intersection of mathematics, biology, and anthropology. As such, it has two basic goals. First, the book provides an *introduction* to the study of population genetics, which provides the mathematical basis of evolutionary theory by describing changes in the frequency of genetic variants from one generation to the next. Second, this introduction has been designed for specific application to *human* populations. Although population genetics is a field that applies to all organisms, the focus throughout this book, particularly in case studies, is on human populations. As an anthropologist, my interest is by definition primarily on *human* populations and genetic diversity. Not that this book has no utility outside of human populations—far from it. I have designed this book to provide a simple introduction to population genetics with minimal mathematics that can be used by advanced undergraduate and graduate students in a variety of fields, including anthropology, biology, and ecology. If you are using this book in one of those other disciplines, rest assured that the same basic principles presented here are applicable to organisms, and your instructor will likely provide other, nonhuman, case studies for clarification. You need not have a detailed background in genetics, although this book is intended for students that have had some initial grounding in genetics, such as one would obtain from an introductory course in biological anthropology or biology.

FORMAT AND ORGANIZATION OF THE BOOK

A quick look through the pages of this book will reveal a number of formulas. This may seem intimidating, but it is not. Although some elementary mathematics is needed to understand population genetics, we do not have to use very advanced math to learn the basics. Throughout this book, we will use only simple algebra of the type that you likely learned in middle or senior high school and some basic concepts of probability, which are developed in the text as we proceed. I also use additional ways, beyond equations, to present the material. Although it is a wonderful experience to glance at a mathematical formula and gain immediate insight into what that formula says about reality, it is (at least for me) a rare experience. I usually have to look at a graphic representation of the formula or utilize an analogy to understand the underlying ideas. Thus, this text uses a lot of graphs and analogies to make the basic points and help you relate the evolutionary process to mathematical ideas.

As with any field, population genetics has its own set of terms. Anything specific to genetics or population genetics is defined in the text, with an additional glossary at the end of the book collecting all such terms. All glossary terms are marked in **boldface** in the text the first time they appear. In-text citation is used in this text, where specific citations are references by author(s) name(s) and year, such as ''Relethford (2004).''

ACKNOWLEDGMENTS

I owe much thanks to Matt Cartmill and Kaye Brown, series editors of the Wiley-Blackwell *Foundation of Human Biology* series, for inviting me to write this book, and for their careful analysis and discussion of the book's goals and structure. I am also very grateful for the guidance and advice of my editor, Karen Chambers. She was a delight to work with on this project. Thanks also to Anna Ehler, Editorial Assistant, and Rosalyn Farkas, Production Editor, for all of their help and attention to my constant questions.

I was first introduced to the study of population genetics in 1975 when I met my graduate school advisor, Frances Lees. I owe Frank a lot for his guidance and friendship over the years in addition to his patience at teaching me population genetics. He got me started both in my profession and in this particular field. I am also very grateful to his academic advisor, Michael Crawford, for helping me learn even more about population genetics over the course of several decades of friendship and collaboration on research projects.

I have worked with other colleagues on research in human population genetics. Two of these colleagues stand out in particular—John Blangero and Henry Harpending. My work with them has been a high point of my career.

Looking back, I can identify many friends and colleagues over the years with whom I have shared discussions at some level or another on population genetics. Some of these have been coauthors, and others have been colleagues with similar interests who have shared one or many conversations or emails. They all have contributed to my understanding of human population genetics. Needless to say, my errors are mine and mine alone. This is the list (and my most sincere apologies if I have missed anyone): Guido Barbujani, Deborah Bolnick, the late Ellen Brennan, Ranajit Chakraborty, Ric Devor, Ravi Duggarali, Elise Eller, Alan Fix, Jon Friedlaender, Rosalind Harding, Mike Hammer, John Hawks, Jeff Heilveil, Keith Hunley, Cashell Jaquish, Lynn Jorde, Lyle Konigsberg, Tibor Koertvelyessy, Ken Korey, the late Gabe Lasker, Paul Leslie, Jeff Long, Lorena Madrigal, Andrea Manica, Yoshiro Matsuo, Jim Mielke, Andy Merriwether, John Mitchell, Kari North, Carolyn Olsen, Esteban Parra, Alan Rogers, Charles Roseman, Dennis O'Rourke, Lisa Sattenspiel, Michael Schillaci, Tad Schurr, Steve Sherry, Peter Smouse, Bob Sokal, Dawnie Steadman, Anne Stone, Mark Stoneking, Alan Swedlund, Alan Templeton, Forrest Tierson, John VandeBerg, Noreen von

Cramon-Taubadel, Tim Weaver, Ken Weiss, Dick Wilkinson, Sarah Williams-Blangero, Milford Wolpoff and Jim Wood. Special thanks to Alan Bittles for providing me with references on inbreeding. I also acknowledge my debt to three individuals whom I have never met, but have spent many hours studying their insightful writings: Luca Cavalli-Sforza, Newton Morton, and the late Sewall Wright.

Last, but certainly not least, I dedicate this book to the five people who mean the most to me in the world—my wife, Hollie Jaffe; my sons, David, Ben, and Zane; and my mother-in-law, Terry Adler. Thanks to all for putting up with me and loving me.

JOHN H. RELETHFORD
State University of New York

CHAPTER 1

GENETIC, MATHEMATICAL, AND ANTHROPOLOGICAL BACKGROUND

My interest in human population genetics started with my difficulty in picking a major in college.

As is often the case, my interests as an undergraduate student were varied, including fields as different as sociology, biology, geography, history, and mathematics. Each of these fields appealed to me in some ways initially, but none sufficiently to take the 10 or more courses to complete an academic major. As I shifted almost daily in my search for a major, I stumbled across anthropology, a discipline that is characterized by academic breadth across the liberal arts. In the United States, anthropology departments are most often constructed around the four-field approach championed by the famous early twentieth-century anthropologist, Franz Boas. Here, anthropology is divided into four subfields: (1) *cultural anthropology*, which examines behaviors in current and recent human populations; (2) *archaeology*, which reconstructs cultural behavior in prehistoric and historic human societies; (3) *linguistics*, the study of language, a uniquely human form of communicating culture; and (4) *biological anthropology* (also known as *physical anthropology*), which focuses on the biological evolution and variation of the human species.

With its focus on both cultural and biological aspects of humanity, and its concern with natural science, social science, and the humanities, anthropology proved to be the perfect liberal arts major for someone like me, who had a difficult time picking any single major. Over time, however, I found myself gravitating more toward the subfield of biological anthropology as I became fascinated by the ways in which humanity had evolved. As I entered graduate school, I wound up concentrating more and more on the nature of human biological variation, and questions about our species' biological diversity. How are human populations similar to and different from each other biologically? How do these differences relate to the process of evolution, and how do these processes relate to human history, culture, and the environment? In one form or another, these

Human Population Genetics, First Edition. John H. Relethford.
© 2012 Wiley-Blackwell. Published 2012 by John Wiley & Sons, Inc.

questions have been at the root of many of the research topics I have focused on during my career, ranging from the effect of historical invasions on genetic diversity in Ireland, to changing patterns of marriage and migration in colonial Massachusetts, to the effect of history and geography on cranial shape across the world.

Underlying all of these questions is the subject of this book, human population genetics, which is a field that has the same breadth of topics that guided my search for a college major. Although this book focuses on *human* population genetics, it is important to realize that population genetics is a subject that concerns *all* organisms. Much of this book consists in explaining basic principles of population genetics, applicable to many species, with further illustration describing case studies from human populations. If you are reading this book in a course on general population genetics, as is often taught in biology departments, for example, you are likely to encounter further case studies on a variety of other species.

I. THE SCOPE OF POPULATION GENETICS

Before getting too far into the application of population genetics to the human species, it is useful to answer the basic question "What is population genetics?" This question can be answered by considering the nature of the broader field of genetics, the study of heredity in organisms. Genetics can be studied at various levels. The study of molecular genetics deals with the biochemical nature of heredity, specifically DNA and RNA. At this level, geneticists focus on the biochemical nature of heredity, including the structure and function of genes and other DNA sequences.

The study of Mendelian genetics, named after the Austrian monk, Gregor Mendel (1822–1884), is concerned with the process and pattern of genetic inheritance from parents to offspring. Mendel's work gave us a basic understanding of how inheritance works, and how discrete units of inheritance combine to produce genotypes and phenotypes. Whereas the focus of molecular genetics is on the transmission of information from cell to cell, Mendelian genetics focuses on the transmission of genetic information from one individual (a parent) to another (the offspring). Mendelian genetics is in essence a statistical subject, dealing with the probability of different genotypes and phenotypes in offspring. A classic example concerns two parents, each of which carries one copy of a recessive gene. The principles of probability show that the chance of any given offspring having *two* copies of that gene, one from each parent, is $\frac{1}{4}$. These principles will be reviewed later, but for now, you should just consider that the transmission of genetic information is subject to the laws of probability.

Population genetics takes this concern with the probability of transmitting genetic information from one generation to the next and extends it to the next level, an entire population (or set of populations, or even an entire species). In population genetics, we are concerned with the genetic composition of the entire population, and how this composition can change over time. For example, consider the classic example of the peppered moth in England. This species of moth comes in two forms, a dark-colored form and a light-colored form. Centuries ago, most moths were light-colored, and only about 1% were dark-colored. Dark-colored moths were rare because they would be more clearly visible against the light color of the tree trunks, making it easier for birds to see them and eat them. Over time, the environment changed, and the frequency of dark-colored moths increased as the frequency of the light-colored moths decreased. Because the color of the moths reflects genetic differences, this observed change is an example of the genetic composition of a species changing over time. Population genetics deals with explaining such changes. In this case, the initial origin of a different form is due to mutation, and the change in moth color over time reflects natural selection, because the environment had shifted following the Industrial Revolution, leading to darker tree trunks, thus creating a situation where dark-colored moths were less likely to be eaten by birds.

When the genetic makeup of a population changes over time, even in a single generation, we have a case of evolution. Population genetics is the branch of genetics that deals with evolutionary change in populations of organisms, and provides the mathematical basis of evolutionary theory. Note that I am using the word *theory* here in the context of the natural sciences, where a theory is a set of hypotheses that have been tested and have withstood the test of time, as compared with the popular use of the word theory as a simple hypothesis. When we speak of evolutionary theory, we are *not* stating that evolution may or may not exist, but instead are referring to a set of principles that explain the *facts* of evolution (in other words, beware of the statement that "evolution is a theory and not a fact," because it is actually *both* a fact and a theory).

Evolution can be viewed over different scales of time and units of analysis. Population genetics deals with changes within a species over relatively short intervals of time, typically on the order of a small number of generations. This type of evolutionary change is also known as **microevolution**, and is contrasted with **macroevolution**, which focuses on the evolution of species and higher levels (genera, families, etc.), and typically deals with geological timescales, ranging from thousands to millions of years. Although macroevolution and microevolution are related in a theoretical sense, there is continued debate over the extent to which long-term macroevolutionary events are a straightforward extrapolation of microevolutionary trends (Simons 2002). The focus of this book is primarily on the theory of microevolution.

Population genetics is concerned with changes in genetic variation over time, that is, genetic differences and similarities. There are two ways of looking at genetic variation: variation *within* populations and variation *between* populations. The former refers to differences and similarities of individuals within a population; the latter refers to average differences between two or more populations. Later chapters will introduce quantitative measures of within-group and between-group variation based on genetic traits, but for the moment, I will use a simple analogy looking at adult human height. Picture yourself in a large classroom filled with students, and imagine that we measured everyone's height. We would use these measurements to compute how much variation existed *within* the classroom. If, for example, everyone in the class were of exactly the same height, there would be *no* variation. If, however, there were differences in height, with everyone being between 5 ft 8 in tall and 5 ft 10 in. tall, then variation would exist because not everyone would be the same. If everyone were between 5 and 6 ft tall, there would be even more variation.

On the other hand, suppose that we want to compare the height in your classroom with the height in the next classroom. An example would be if the average height in your classroom were 5 ft 9 in. and the average height in the other classroom were 5 ft 8 in. The difference in average height would be 1 in. This difference would be an example of variation *between* groups. If the average height of the two classes were the same, then there would be no variation between groups. In evolutionary terms, we are interested in changes in genetic variation that take place both within and between populations.

By studying genetic change over time and its effects on genetic variation within and between populations, we are able to apply the theory of population genetics to address a wide variety of questions about human variation and evolution. A small sample of such questions (which will be addressed in later chapters) includes

- How much inbreeding occurs in human populations, and what is the effect of this inbreeding?
- What does genetic variation tell us about our species' history?
- Can genetics to be used to trace ancient human migrations?
- Where did the first Americans come from?
- Why do some human populations have high frequencies of the harmful sickle cell allele?
- Are certain genes resistant to acquired immunodeficiency syndrome (AIDS)?
- Why do some small populations differ genetically from their neighbors to such an extent?
- What impact does geography have on our choice of mates?

Even this short list shows that population genetics has relevance to many questions about human biological variation and evolution. In addition, the general principles of population genetics are used to address the same concerns—variation and evolution—in all organisms. In short, population genetics is a key to understanding life. Although this book focuses on human populations (because of my interests and training), never forget that many of the general principles of population genetics apply across the span of life itself.

As noted earlier, the study of human population genetics examines the application of mathematical principles and models to the transmission of genetic information from one generation to the next in human populations. Population genetics can be regarded here as a field that combines genetics, mathematics (especially probability), and anthropology. The remainder of this introductory chapter provides a brief review of some basic principles of genetics and probability, and concludes with a broader consideration of how population genetics applies in an anthropological context.

II. GENETICS BACKGROUND

Considering the nature of this book and its intended audience, one might assume that you are a student in a course on population genetics or a related field. Typically, such students have had some background in some basic concepts of genetics, particularly Mendelian genetics, from high school as well as in an introductory college course in biology or biological anthropology. As such, the following information is not meant to be a detailed discussion of genetics, but instead a brief review of some high points and terminology in order to dive into population genetics as quickly as possible. More detail will be given as needed throughout the text. If you find that the following brief review is a bit too brief, I suggest getting more review and/or detail from comprehensive Internet sources such as *Wikipedia*, browsing through some introductory genetics books, and consulting with your professor.

Most discussions of genetics start with mention of deoxyribonucleic acid (DNA), often referred to casually as "the genetic code." Although we are learning more every day about the nature of DNA and how it works, many of the basic principles of population genetics were derived long before much was known about DNA. Indeed, James Watson and Francis Crick discovered the biochemical structure of DNA in 1953, whereas many ideas in population genetics were first developed in the 1930s and 1940s. Although advances in molecular genetics have certainly affected continued development of population genetics in terms of both theory and methods (as will be described later), many of the basic concepts of genetic transmission in populations were developed before we really knew the structure and function of exactly what was being transmitted.

The DNA molecule is made up of two strands that consist of nucleotides, molecules that contain a nitrogen base connected to sugar and phosphate groups. There are four different bases in DNA: adenine (A), thymine (T), cytosine (C), and guanine (G). The sequence of these four different bases make up the genetic "code," and by analogy they can be considered "letters" in a four-letter DNA alphabet. A related molecule, ribonucleic acid (RNA), is involved in the transcription of proteins, expression of genes, and other vital biochemical functions. A critical aspect of DNA is that the A and T bases pair up as do the C and G bases. As DNA is double-stranded, this means that an A on one strand is paired with a T on the other strand. Likewise, T is paired with A, C with G, and G with C. This property of DNA allows it to make copies of itself, thus ensuring the transmission of genetic information from cell to cell. The pairing of bases between the two stands is known as a **base pair** (abbreviated bp), and the length of DNA sequences is measured by the number of base pairs.

A. Mendel's Laws

Much (though not all) of our DNA exists on long strands in the nuclei of our cells, called **chromosomes**. Chromosomes come in pairs. Different species have different numbers of chromosomes; humans have 23 pairs, whereas chimpanzees (our closest living relative) have 24 pairs. During the replication of body cells through **mitosis**, a single cell containing 23 pairs of chromosomes will duplicate, giving rise to two identical cells, each with 23 pairs of chromosomes. However, this is not what happens during reproduction. Instead of passing along 23 pairs of chromosomes to your offspring in a sex cell (sperm in males, egg in females), you pass on *one* of each pair through the process of **meiosis**. The process of chromosome pairs separating through meiosis is also known as **Mendel's law of segregation** (or, sometimes, as Mendel's first law). You contribute 23 chromosomes (but not 23 *pairs*), and your mate contributes 23 chromosomes, resulting in your child having $23 + 23 = 23$ chromosome pairs. Likewise, your genetic inheritance also resulted from this process, as one of each chromosome pair came from your mother and the other one came from your father.

As a bisexual organism (a species that has two distinct sexes, male and female), half of your genetic inheritance comes from your mother and half from your father. The same applies to any biological siblings. Apart from identical twins, why are you not genetically identical to a sibling? If my brother and I both received 50% of our DNA from our mother and 50% from our father, why are we not genetically the same? The answer relates to basic probability; we do not inherit the *same* 50%. For any given chromosome pair, there is a 50 : 50 chance of one being passed on to an offspring, either the maternal chromosome (from your mother) or the paternal chromosome (from your father). For example, imagine that I have passed along my maternal chromosome for the first chromosome pair to a child. The next child may or may not receive the same maternal chromosome; it is a 50 : 50 chance for either the maternal or the paternal chromosome. The same probability applies to each chromosome pair, as they are all *independent* such that whatever chromosome you pass on from the first chromosome pair has no effect on the second pair, the third, and so on.

We can illustrate this principle with a simple analogy using coins. Imagine an organism with only three chromosome pairs, each represented by a penny with two sides—heads and tails. If we flip the first coin, we have a 50 : 50 chance of getting heads (H) or tails (T). We will use this as a model for a chromosome pair consisting of one chromosome labeled H and one labeled T. If you flip heads for the first coin (chromosome pair), what is the probability of flipping heads on the *second* coin? It is still 50 : 50 because the coin flips are independent; the outcome of one coin flip does not influence any other coin flips. In terms of the genetic analogy, this hypothetical organism can produce eight different combinations of coin flips. One of these eight combinations would be getting heads for the first coin, heads for the second coin, and heads for the third coin. Another possibility would be heads for the first coin, heads for the second coin, and tails for the third coin. If we follow this pattern, we wind up with eight different combinations, each equally likely:

1. Heads–heads–heads
2. Heads–heads–tails
3. Heads–tails–heads
4. Heads–tails–tails
5. Tails–heads–heads
6. Tails–heads–tails
7. Tails–tails–heads
8. Tails–tails–tails

Because of chance, this organism could produce eight different combinations of chromosomes. This independent inheritance is known as **Mendel's law of independent assortment** (or Mendel's second law).

In principle, we could simulate the same process for human beings by using 23 different coins, but it take much too long to enumerate all possible combinations of coin flips. Instead, we can figure out the number of possibilities using the simple formula 2^n, where n is the number of coins/chromosome pairs. For humans, $n = 23$ chromosome pairs, giving $2^{23} = 8,333,608$ combinations! Keep in mind that this is for one individual. The same rule applies to the production of sex cells in the individual's mate; they, too, can produce up to 8,388,608 combinations. A child could therefore have any of the first parent's combinations paired with any of the second parent's combinations, giving a total of $8,388,608 \times 8,388,608 = 70,368,744,177,664$ possible genetic combinations in any given child! Given the number of possibilities, it is easy to see why it would be virtually impossible for me to be genetically identical to my nontwin brother for my entire **genome**.

As is typically the case when explaining basic models of reality, I have to point out that all of the above is actually a bit of an oversimplification. The basic process is further complicated by **recombination**, which involves the crossover of sections of DNA of chromosome pairs during meiosis. Start with a pair of chromosomes, with one chromosome from the mother and one from the father. During meiosis, the pair does no segregate exactly, such that pieces of the mother's DNA are exchanged with pieces of the father's DNA. Thus, any sex cell that you pass on to an offspring is unlikely to follow the ideal Mendelian model of being either your mother's chromosome or your father's chromosome, but instead reflects parts of both. The process of recombination provides even more shuffling of genetic combinations with each generation.

Through meiosis with recombination, a new generation can reflect different combinations of what was present in the parental generation. However, in terms of the overall genetic composition of the population (how many different genetic forms exist), this reshuffling does not change anything. An analogy here would be a deck of cards. Each time you shuffle the deck and deal out a five-card poker hand, you are likely to get a different combination, such as a three of clubs, five of spades, six of spades, ten of hearts, and a queen of diamonds. Return these cards to the deck, shuffle, and deal again. You are most likely to have a completely different hand (it is possible to get the same hand, but extremely unlikely, as there are 2,598,960 possible different five-card poker hands using 52 cards and no jokers). Each time you shuffle and deal, you can get a new combination, but the basic composition of the deck has not changed—you still have four suits each with 13 cards ranging from 2 through ace. Nothing new would happen unless there were a *mutation* in the deck, say, resulting from changing a 10 of spades to a brand new type of card, such as an 11 of spades. (Don't try this in a real game!) Population genetics involves understanding how the genetic composition of a population can change through the operation of mutation and other forces of evolution.

B.   Alleles, Genotypes, and Phenotypes

What is a **gene**? As with many core ideas and concepts (e.g., life, love, culture, race), the actual definition of gene has changed over time and is often difficult to pin down (Marks and Lyles 1994). The term *gene* was first used in a very general way to refer to a *unit of inheritance*. With the growth of molecular genetics, it has become more common to refer to a gene in a more specific sense, which is a DNA sequence associated with a functional product, such as a protein. This more restricted definition does not include noncoding sections of DNA. Although some population geneticists use the more current restricted definition (e.g., Hamilton 2009), others use the more general definition for convenience (e.g., Hedrick 2005). Here, I will use the more specific restricted definition to comply with your likely background in genetics, and refer to the entire genome as consisting of genes and other DNA sequences. The broader term **genetic marker** is often used to refer to any gene or DNA sequence that has a known location on a specific chromosome.

When we study a genetic marker, we refer to its specific location on a particular chromosome; this location is referred to as a **locus** (plural **loci**). A key concept in population genetics is the **allele**, which refers to alternative forms of a gene or DNA sequence at a given locus. Loci that have two or more alleles that are not rare (typically defined as a frequency greater than 0.01) are called **polymorphisms**, which literally translates as "many forms."

As an example of the concept of allele, consider the gene that affects lactase production in humans. As mammals, humans rely on milk during infancy. We produce an enzyme (lactase) in order to break down milk sugar (lactose). A specific gene (LCT) is located on chromosome 2 and regulates the production of lactase. There are several different forms (alleles) of this gene. One allele (R) causes enzyme production to decrease during early childhood (an age by which humans have been weaned), and another allele (P) allows continued high production of lactase into adulthood, a condition known as *lactase persistence*. There is also a third rare allele, but it will not be discussed in this example (Mielke et al. 2011).

For any trait in your nuclear DNA, including lactase activity, you inherit two copies of the gene or DNA sequence, one from your mother and one from your father, which collectively makes up your **genotype**. In the case of lactase activity, there are two main alleles (R and P) in the human species, which means that there are three possible genotypes. Some individuals will inherit two copies of the P allele and will have the genotype PP, while others will inherit two copies of the R allele and have the RR genotype. Both people with PP and RR genotypes are **homozygous** for this trait, which means that they have inherited the same allele from both mother and father. There is a third possibility, which is the genotype PR, where the person has inherited a P allele from one parent and an R allele from another parent (it does not matter which parent gave the P allele and which gave the R allele). When someone inherits a different allele from each parent, that person is **heterozygous** for that trait.

What are the different outcomes for these different genotypes? Each has inherited genetic information regarding the restriction or persistence of lactase production. The physical manifestation of a genotype is known as the **phenotype**. In complex traits, such as height or skin color, the phenotype is a reflection of the genotypes of the different genes that affect the trait as well as environmental effects, such as nutrition in the case of height, or solar exposure in the case of skin color. In "simple" genetic traits, such as lactase activity, the phenotype is determined by the genotype and which, if any, alleles are dominant or recessive. The effect of a **dominant** allele is noticeable even if only one copy is present, whereas a **recessive** allele's effect can be masked by a dominant allele. In the case of lactase activity, the P allele (lactase persistence) is dominant and the R allele (restriction) is recessive. This means that someone who inherits one *or* two P alleles will show the lactase persistence phenotype, and those that inherit two R alleles will show lactase restriction. In other words, lactase persistence can result from either the PP or PR genotypes, and lactase restriction can result only from having the RR genotype.

It is important to remember that *dominant* and *recessive* refer to the nature of the alleles and have nothing to do with the actual frequency of an allele; that is, a dominant allele is not necessarily more common than a recessive allele. For example, in humans there is a condition resulting in extra fingers or toes (polydactyly) that is caused by a dominant allele, yet it is very rare in occurrence (Wolf and Myrianthopoulos 1973). Another example in humans is the ABO blood group, where the most common allele in our species is the O allele, which is recessive.

For any given locus, the alleles need not be either dominant or recessive. For many loci, the alleles are **codominant**, meaning that the effect of both alleles is expressed in the phenotype. An example in humans is the MN blood group located on chromosome 4, which has two alleles, M and N, which produce different molecules on the surface of our red blood cells; the M allele produces type M molecules and the N allele produces type N molecules. Given these two alleles, we have three possible genotypes: MM, MN, and NN. What about the phenotypes? Logically, we can see that the homozygous genotype MM will result in type M molecules because both alleles contain the same message—type M blood. It is also clear that the genotype NN will produce type N molecules. What of the genotype MN? The phenotype associated with a heterozygous genotype depends on whether one allele was dominant. In this case, the M and N alleles are codominant, which means that *both* the M allele and the N allele will manifest, resulting in the production of *both* type M and type N molecules. In the case of a codominant locus, each genotype has a distinct phenotype. As we will see in Chapter 2, this makes it much easier to count alleles and determine their frequency (a vital part of population genetics).

Before moving on, I want to point out some other complications. Although most examples in this book use a simple model of a single locus with two alleles, in reality there are actually many loci with more than two alleles, and some loci where there are dozens of alleles. Basic concepts will be introduced using the simple two-allele model where possible and bringing in this additional complication where appropriate.

Another complication is the fact that some loci have dominant, recessive, *and* codominant alleles. A good example for humans is the ABO blood group, located on chromosome 9. There are three main alleles, A, B, and O, where the A allele codes for type A molecules, the B allele codes for type B molecules, and the O allele codes for neither of these. In the ABO system, the *O* allele is recessive and the *A* and *B* alleles are codominant. Given three possible alleles, there are six possible genotypes: *AA, BB, OO, AO, BO*, and *AB*. What are the possible phenotypes?

The phenotypes of the three homozygous genotypes (*AA, BB, and OO*) are easy to determine. Genotype *AA* produces type A blood, genotype *BB* produces type B blood, and genotype *OO* produces type O blood. The phenotypes of the three remaining genotypes can be determined by knowing which alleles are dominant, recessive, or codominant. Because the *O* allele is recessive, those with genotype *AO* will show only the effect of the dominant *A* allele, and hence will have type A blood. Likewise, those with genotype *BO* will show only the effect of the dominant *B* allele, and will have type B blood. The remaining genotype, *AB*, has two codominant alleles, which means that both A and B molecules will be produced, and people with this genotype therefore have what we call type AB blood. For the ABO blood group, there are three alleles that can form six different genotypes that correspond to four different phenotypes (ignoring for the moment additional complications, such as the fact that there are actually two subtypes of the *A* allele).

C.   How Do We Assess Human Genetic Diversity?

As will be clear in later chapters, much of the core of population genetics theory is abstract, dealing with hypothetical alleles at hypothetical loci in hypothetical populations. Although hypothetical rumination is interesting in and of itself, the ultimate test of a mathematical model of reality is to see how well it represents reality, which means that at some point we need information about real alleles at real loci in real populations! Although a variety of loci and traits will be provided in case studies throughout this book, it is useful to look briefly at some of the different ways anthropologists and geneticists use to assess genetic diversity.

Red Blood Cell Markers

For the first half of the twentieth century, most information on genetic diversity in human populations came from the study of blood types based on red blood cell groups, where phenotypes were based on the reaction of antigens present in the blood with corresponding antibodies (Boyd 1950). In the ABO blood group system, for example, this is based on reactions of A and B antigens with their respective antibodies, anti-A and anti-B. Suppose that someone's blood shows a reaction with the anti-A antibody but not the anti-B antibody. This means that they have the A antigen but not the B antigen, and therefore have blood type A, and therefore either the *AA* or the *AO* genotype. There are many different red blood cell systems, including ABO, Rhesus, MN, Kell, Diego, Duffy, and P (Crawford 1973).

By the 1960s and 1970s, technological advances such as electrophoresis had led to a proliferation of other genetic markers of the blood. **Electrophoresis** involves passing an electric current through a gel. Blood samples are placed at the negative pole of the gel and, as current flows from negative to positive, molecules move through the gel. Because different molecular structures move at different rates, the process allows identification of different molecular structures associated with different genotypes. Applied to blood samples from anthropological surveys, a vast amount of data were collected on numerous red blood cell protein and enzyme loci (Crawford 1973; Roychoudhury and Nei 1988; Cavalli-Sforza et al. 1994).

DNA Markers

Genetic markers of the blood, including markers based on white blood cells, are now labeled as *classical genetic markers*, contrasted with the newer DNA markers. Although classical markers provide information on genetic variation, DNA markers provide a closer window on genetic variation, moving beyond the level of molecular variability to the underlying level of DNA variation.

One method of DNA analysis involves the identification of **restriction fragment length polymorphisms** (RFLPs). Restriction enzymes that are produced by different types of bacteria can bind to sections of a DNA sequence and cut that sequence at a particular point. For example, the *EcoRI* bacterial enzyme will bind to the 6-bp sequence GAATCC (which, by definition, corresponds to the sequence CTTAGG on the other DNA strand). If this sequence is present in a DNA sample, *EcoRI* will cut the sequence between the G and the first A, producing two fragments, one with the base G and the other with the sequence AATCC. If the DNA sample did not contain the sequence GAATCC, but instead had a mutation resulting in GATTCC (where the second A mutated into T), then the target sequence would not be recognized and the DNA sample would not be cut. Depending on the presence or absence of certain DNA sequences, a DNA sample might be cut into fragments of different lengths.

Another type of DNA variation widely studied in human populations consists of repeated DNA sequences, such as CACACACACACA, where the 2-bp sequence is repeated 7 times. Because of mutation, the number of repeats can go up or down, resulting in variation. **Short tandem repeats** (STRs), also known as *microsatellite DNA*, are widely used in studies of human populations. STRs consist of short repeated sequences consisting of 2–5 bp. Longer repeated sequences, known as *minisatellites*, are also used.

Another form of DNA analysis looks for **single-nucleotide polymorphisms** (SNPs), where DNA sequences differ by one base, such as having the base C on one sequence versus the base T on another sequence:

> Sequence 1: TATTCCGGA
> Sequence 2: TACTCCGGA

In this case, the two sequences differ at the third position, and there are two alleles: the first has the base T and the second has the base C. SNP variation is being increasingly studied in human populations; as of 2007, over 3.1 million SNPs had been identified (International HapMap Consortium 2007).

Haplotypes

Loci that are close together on the same chromosome tend to be inherited together (**linkage**). A **haplotype** is a combination of alleles that are inherited as a single unit. Haplotypes are sometimes defined as a set of linked loci, and can be based on RFLPs, STRs, SNPs, or combinations of these. For example, Foster et al. (1998)

conducted a genetic analysis on descendants of male relatives of the third US president, Thomas Jefferson, to see if there was a genetic connection between this family and the descendants of Eston Hemings, son of Sally Hemings, who was an enslaved African-American woman. This study involved using a haplotype of the Y chromosome that was unique to the Jefferson family. This haplotype consisted of seven SNPs, 11 STRs, and 1 minisatellite.

Mitochondrial DNA and Y-Chromosome DNA

The discussion so far has dealt with nuclear DNA and inheritance from both parents, the traditional way to present Mendelian genetics. These examples refer to **diploid** inheritance (two copies). Although the majority of examples presented in this text refer to diploid inheritance, some traits in humans are **haploid**, and come from only one parent. One type of haploid inheritance is **mitochondrial DNA** (abbreviated mtDNA). Although most of our DNA is contained in the chromosomes, there is a small amount in the mitochondria, which are the cell structures responsible for energy production. In humans, mitochondrial DNA, the circular DNA molecule is typically 16,569 base pairs in length, which is a very small fraction of the more than 3 billion base pairs of nuclear DNA.

What makes mitochondrial DNA so fascinating and useful is the way it is inherited. Unlike nuclear DNA, which is inherited from both parents, you inherit mitochondrial DNA only from your mother. This pattern of inheritance results from the way in which sperm and egg combine to form a zygote (fertilized egg); the mitochondria in the zygote comes from the egg, and thus contains only the mother's genetic contribution. Transmission from one generation to the next is through the female line. Although males inherit mitochondrial DNA from their mothers, they cannot pass it on; transmission occurs only in the female line.

This exclusive maternal inheritance simplifies genealogical analysis, figuring out where certain alleles came from. With nuclear DNA, it is difficult to do this because the number of potential ancestors doubles with each generation in the past—you have two parents, four grandparents, eight great grandparents, and so on. The number of potential ancestors and recombination every generation make it difficult to tell if a particular allele came from any given ancestor. With mitochondrial DNA, you have only one ancestor in any generation in the past. One generation back, that ancestor would be your mother; two generations back would be your mother's mother, and so on into the past. Another advantage of mitochondrial DNA is its use in ancient DNA analysis because the high number of copies per cell means it that will be more likely to survive degradation compared with nuclear DNA (O'Rourke 2007). Mitochondrial DNA is also used in ancestry testing, a service available from a number of vendors. The problem here is that such tests can tell you about ancestry in only one line. For example, you may have a maximum of 16 great-great grandparents (see Chapter 3 for why this is a maximum and how you can actually have fewer great-great grandparents). Mitochondrial DNA analysis will tell you about only one of these 16 ancestors—your mother's mother's mother's mother—and not your mother's mother's mother's father, your mother's mother's father's mother, or any of your other ancestors.

Just as ancestry can be traced in the maternal line using mitochondrial DNA, a similar pattern of ancestry is found in **Y-chromosome DNA**. Of our 23 pairs of chromosomes, one pair is the sex chromosomes and the other 22 pairs are referred to as **autosomes**. There are two types of sex chromosomes, X and Y, which determine biological sex; females have two X chromosomes, whereas males have one X chromosome and one Y chromosome. Males receive their X chromosome from their mother and their Y chromosome from their father. When sperm are produced, the sex chromosomes segregate and males pass on either their X or their Y. Because there is very little recombination of the Y chromosome with the X chromosome, this means that males pass on their Y chromosome almost intact to their sons. Analysis of the nonrecombining part of the Y chromosome provides the same insights as mitochondrial DNA, although in the father's line.

Quantitative Traits

Although the focus of this book is primarily on the application of population genetics theory to a simple single-locus model, it is useful to point out that there are also extensions to more complex traits, particularly those that involved the effects of both genetics and the environment. Before the dawn of the twentieth century, the only way to approximate genetic variation was through physical traits, such as cranial and facial measurements. The study of **quantitative genetics** deals with such traits. Many physical traits, such as height, head length, and skin color, are examples of **quantitative traits**, whose phenotype varies continuously. For example, if we consider height, and see one person with a height of 5 ft 6 in. and another person with a height of 5 ft 7 in., we know that it is possible to find someone with any intermediate value, such as 5 ft 6.5 in., 5 ft, 6.6 in., and so on.

Quantitative traits are due to the joint influence of one or more loci and environmental influences, where the latter can include a variety of influences ranging from prenatal environment to climate to diet, among many others. Often, quantitative traits are **polygenic**, meaning that two or more loci interact to produce a genotype. In some cases, an **equal and additive effects model** can be used to describe polygenic inheritance, where all of several loci contribute equally to the same phenotype. In other cases, one locus may have a more substantial effect, such as in a **major gene model**. In all cases, the phenotype (e.g., how tall you are, the color of your skin) reflects the joint effect of these loci and environmental influences. Thus, two people could have the same genotypic inheritance but grow to different heights because of exposure to different environmental conditions. Likewise, two people may have different genotypes but wind up with the same phenotype because of environmental influences.

With the advent of biochemical and DNA markers, attention has moved away from quantitative traits. This does not mean, however, that they are without value. Although newer analytic methods have now allowed such traits to provide useful measures of human population relationships and history (Relethford 2007; von Cramon-Taubadel and Weaver 2009), they will only be referenced briefly in this text, with the bulk of attention given to "simpler" genetic traits.

III. PRINCIPLES OF PROBABILITY

Population genetics is mathematical in nature. The use of mathematics can be scary to many students. Indeed, I find some of the more complex methods in the field scary myself! However, the basic concepts of population genetics can be learned with a minimum of mathematics. The level of math used in this text assumes some previous background in the basic algebra learned in high school. I also use many graphs to give a visual feel for the mathematical relationships that may not be directly apparent from the raw equations. Some algebra will be used for proofs, many of which are explained at the end of each chapter to avoid interference with the flow of text. There will be one placed where additional explanation will be appended for those who know a bit of calculus, but that is not necessary for understanding the basic concepts.

Most of what we look at in population genetics to model the evolutionary process consists in applying some basic concepts of probability. If you think about it, a number of questions about reproduction and inheritance boil down to questions about probability. For example, what is the probability that a parent will pass on a given allele to a child? At the population level, similar questions regarding the probability of genetic transmission apply. What is the probability that a given locus will mutate within a generation? What is the probability that a given allele will increase over time, decrease over time, or stay the same? What is the probability that an allele will move from one population to another through migration in a given generation? These and other questions boil down to questions of probability.

A. Some Simple Rules of Probability

Many basic concepts of probability can be demonstrated using coins, dice, and/or a deck of cards. Such everyday objects are useful in learning more abstract concepts because the answers are often more intuitive. For example, consider the very simple questions that can be generated using a single coin; for instance, what is the probability of flipping the coin and getting heads? We all can answer this question immediately—the probability is $\frac{1}{2}$. The same is true for the probability of flipping the coin and getting tails ($= \frac{1}{2}$). This is probably obvious, but where exactly did we get this number? We simply divided the number of times that a specific event (getting a head) could occur by the total number of events that could occur (getting a head or a tail). Consider a different example. Roll a single die, a cube with six numbered sides. What is the probability of rolling the die and getting the number 3? There is only one way to get the number 3, and there are six possible outcomes, so the probability is $\frac{1}{6}$. The same is true of all the other numbers (1, 2, 4, 5, 6); each has a $\frac{1}{6}$ probability. The sum of the probabilities of all possible outcomes is equal to 1. In the case of flipping a coin, there are only two possible outcomes (heads and tails), each with a $\frac{1}{2}$ probability, and the sum of all outcomes is $\frac{1}{2} + \frac{1}{2} = 1$. For rolling a single die, there are six possible outcomes, each with $\frac{1}{6}$ probability, and they add up to 1.

Probabilities are expressed as proportions that can range from 0 to 1. For example, $\frac{1}{6} = 0.167$. Sometimes we here probabilities also expressed as percentages, and all you have to remember is that a percentage is a proportion multiplied by 100. For example, we can say that the probability of rolling a single die and getting the number 3 is $\frac{1}{6} = 0.167$ (a proportion), and we could express this as a percentage (16.7% of the time, we will get the number 3).

In population genetics, we often look at events that are independent of one another, such as Mendel's law of independent assortment. For example, what is the probability of rolling a pair of dice and getting the number 3 on both dice? Again, most of us will tend to solve such problems intuitively. We know the probability of rolling a die and getting the number 3 is $\frac{1}{6}$, and we also know that what comes up on one die in no way influences what comes up on the second die (they are independent). Thus, the probability of getting the number 3 on both dice is $\frac{1}{6} \times \frac{1}{6} = \frac{1}{36} = 0.028$. What we have done here is use what is known as the AND rule in probability. In general, we use the symbols $P(A)$ to refer to the probability that outcome A occurs and the symbol $P(B)$ to refer to the probability that outcome B occurs, such that the probability of both A AND B occurring is their product:

$$P(A \text{ AND } B) = P(A)P(B) \tag{1.1}$$

In formal terms, the question above of rolling a pair of dice and getting two 3s is solved using the AND rule as $P = \frac{1}{6} \times \frac{1}{6} = \frac{1}{36} = 0.028$. As an aside, you might notice that this equation is numbered. All main equations in this text are numbered, consisting of a chapter number (the first number), followed by a period, and followed by a sequence number. Thus, equation (1.1) refers to the first numbered equation in Chapter 1. Main equations are those that are referenced elsewhere in the text.

We are also interested in situations where two outcomes are mutually exclusive, and therefore both cannot be true at the same time. For example, what is the probability of rolling a single die and getting a 3 OR a 4? The probability of rolling a 3 is $\frac{1}{6}$, as is the probability of rolling a 4. These outcomes are mutually exclusive, as you cannot roll a 3 and a 4 at the same time! What then is the probability of A *or* B? The answer uses what is known as the OR rule, where the individual probabilities are added:

$$P(A \text{ or } B) = P(A) + P(B) \tag{1.2}$$

Thus, the probability of rolling a die and getting a 3 or a 4 is $\frac{1}{6} + \frac{1}{6} = \frac{2}{6} = \frac{1}{3} = 0.333$.

Here are a few simple examples for reviewing the basic rules of probability:

1. *Question*: You roll a pair of dice. What is the probability of getting an even number?
   Answer: There are three possible even numbers (2, 4, and 6) out of six possible outcomes. The probability is $\frac{3}{6} = \frac{1}{2} = 0.5$.
2. *Question*: You flip two coins at the same time. What is the probability of having both of the coins come up heads?
   Answer: This is a case that calls for the AND rule, as the answer is the product of each coin coming up heads. The probability is $\frac{1}{2} \times \frac{1}{2} = \frac{1}{4} = 0.25$.

3. *Question*: This is the same problem as 2, but this time, flip *three* coins. What is the probability of all three coming up heads?
   Answer: You still use the AND rule but extend it to three outcomes by multiplying the three probabilities as $\frac{1}{2} \times \frac{1}{2} \times \frac{1}{2} = \frac{1}{8} = 0.125$.

4. *Question*: You have thoroughly shuffled a standard 52-card deck of cards (no jokers!). What is the probability of randomly selecting a card and getting an ace or a face card?
   Answer: You need to use the OR rule to compute the probability. Start with the probability of getting an ace. There are four aces in the deck, so the probability of drawing an ace is $\frac{4}{52}$. There are nine face cards in the deck (three jacks, three queens, and three kings). The probability of getting a face card is therefore $\frac{9}{52}$. Put these together using the OR rule, and the probability of getting an ace or a face card is $\frac{4}{52} + \frac{9}{52} = \frac{13}{52} = \frac{1}{4} = 0.25$.

B. Genetics and Probability

Many problems in Mendelian genetics concern probability. A typical question asks the student to describe the distribution of possible genotypes and phenotypes in the offspring of a given mating. To illustrate this, consider a hypothetical locus that has two alleles, *A* and *a*. With two alleles, there are three possible genotypes: *AA*, *Aa*, and *aa*. What is the probability of two *Aa* parents having an offspring with the genotype *AA*? To answer this, we have to connect the basic idea of inheritance with probability. A child who has the *AA* genotype will have inherited the *A* allele from both parents. The question can now be expressed more specifically: What is the probability that both parents pass on an *A* allele? Both parents have the genotype *Aa*, which means that each parent has a $\frac{1}{2}$ chance of passing on the *A* allele. Given this, we use the 'and' rule and give the answer as $\frac{1}{2} \times \frac{1}{2} = \frac{1}{4} = 0.25$.

What about when the parents have different genotypes? The same principles apply. For example, what is the probability of a man with genotype *AA* and a woman with genotype *Aa* having a child with genotype *Aa*? The child must inherit an *A* allele from one parent and an *a* allele from the other parent. In this case, the father can pass on only the *A* allele because he has the *AA* genotype (and obviously cannot pass on a different allele). Therefore, the mother must pass on the *a* allele. What is the probability of the man passing on an *A* allele *and* the woman passing on the *a* allele? Because the man only has *A* alleles, the probability that he will pass on an *A* allele is $\frac{2}{2} = 1$. The woman has the *Aa* genotype, and therefore the probability of her passing on the *a* allele is $\frac{1}{2}$. Using the AND rule, we get the answer to the original question as $1 \times \frac{1}{2} = \frac{1}{2} = 0.5$. In other words, we expect that half the time this couple will have a child with the *Aa* genotype.

It is often more useful to consider the distribution of all possible outcomes of a given mating. A useful tool is the **Punnett square**, a simple method invented by geneticist Reginald Punnett (1875–1967) and one that you may recall from high school and/or college biology. In a Punnett square, we simply construct a 2×2 table that lists the possible contributions of each parent. For the above example, where the father has the *AA* genotype and the mother has the *Aa* genotype, the Punnett square is as follows:

		From the mother	
		A	a
From the father	A	AA	Aa
	A	AA	Aa

This simple table shows us that the two possible contributions from the father are both the *A* allele, whereas there are two different contributions from the mother; she passes on either the *A* allele or the *a* allele. The table also shows that 2 in 4 times the child is expected to have the *AA* genotype, and 2 of 4 times, the child is expected to have the *Aa* genotype. The probabilities for different genotypes among the offspring are $AA = 0.5, Aa = 0.5$, and $aa = 0.0$.

As another example, what is the distribution of possible genotypes for the case where both the man and the woman have the *Aa* genotype? The Punnett square in this case is

		From the mother	
		A	a
From the father	A	AA	Aa
	a	Aa	aa

Note that the case where the father contributes an *a* allele and the mother an *A* allele is the same as the case where the father contributes an *A* allele and the mother an *a* allele. They both produce the same genotype: *Aa*. The probabilities of getting different genotypes in the offspring are $AA = \frac{1}{4} = 0.25, Aa = \frac{2}{4} = \frac{1}{2} = 0.50$, and $aa = \frac{1}{4} = 0.25$.

Keep in mind that these probabilities are just that—probabilities—and not certainties. If this couple has four children, they will not necessarily have one child with *AA*, two children with *Aa*, and one child with *aa*. With four children, they might produce four children with *Aa*, or three with *AA* and one with *aa*, or a number of other possible combinations. The genotype of each child is independent of the others, and with a small number of actual outcomes, we are likely to see considerable variation from the expected numbers. This is known as **genetic drift**, and will be discussed in more detail in Chapter 5.

IV. THE ANTHROPOLOGICAL CONNECTION

When I describe the nature of my research (or the writing of this book), I focus on the mixture of genetics, probability, and anthropology. Most people are familiar with the basic ideas of Mendelian genetics, so the connection between genetics and probability is clear. The connection between anthropology, genetics, and probability may not be immediately clear, but on some reflection makes perfect sense.

A.   What Is a Population?

What exactly is meant in population genetics by the term **population**? The term can actually have a number of different meanings depending on context and on intended use. One useful definition has been offered by Hedrick as a "group of interbreeding individuals that exist together in time and space" (2005:62–63). In this sense, we are focusing on what is sometimes referred to as the *breeding population*, a group from within which mates are typically chosen.

The actual delineation of a population, particularly in humans, is often a bit arbitrary. How much mating must take place within a set of boundaries in order to qualify as a breeding population? Ninety percent? Seventy-five percent? Can a small and culturally homogenous village on a remote island be considered a population? Most certainly. What about a large and diverse city, such as London or New York City? Probably not, because such cities actually encompass a number of smaller subpopulations on the basis of ethnicity, social class, and so on.

In practice, geography is a usual starting point in defining a human population, using geopolitical units such as towns and villages as our operational units of analysis. Sometimes data may not be available with the use of such units, and we might have to consider larger units of analysis, such as townships or territories. If the local geographic units are not culturally homogeneous, we might want to consider further subdividing the sample into smaller units that are more likely to reflect actual mating behaviors. In humans, we might want to consider the impact of a number of cultural factors on mating and, hence, population definition, such as ethnicity, religion, social class, education, and linguistics.

Often, the purpose of the study and consideration of sample size will dictate the definition of the population. In my own research in Ireland, for example, I have used both broad and local geopolitical units for analysis. The island of Ireland, made up today of the Republic of Ireland and Northern Ireland, is divided into four large areas called *provinces*, each broken down into a number of local governmental units known as *counties*. There are 26 counties in the Republic of Ireland. Six counties in Northern Ireland are recognized geographically, but not for administrative purposes. One of the datasets that I have worked with consisted of a number of phenotypic measures of the head and body. For one study, where the purpose was to identify very broad patterns of variation, I used the county for the definition of the "population," even though the patterns did not correspond exactly to the concept as used in many population genetic models (Relethford and Crawford 1995; Relethford et al. 1997). In another study, where the purpose was to look at *local* patterns of variation, I used the town of residence of each individual (Relethford 2008b). In this case, the unit of analysis corresponded more closely to the idealized definition of population at the cost of reduced sample size. Often there is a tradeoff between the number of populations and the sample size per population, and the investigator may have to look at the data from several different levels. The different definitions of population will be highlighted in a number of the case studies presented in this book.

B. Anthropology and Population Genetics

As will be discussed in detail later, the genetic composition of a population at any point in time is a reflection of mating systems and the forces of evolution. In brief, patterns of mating include mating with close relatives (**inbreeding**) and mate preference on the basis of phenotypic preferences (**assortative mating**, such as occurs when individuals choose mates who have similar characteristics). Patterns of mating affect the distribution of different genotypes in a population, and it is clear that human behavior allows for a variety of different effects due to inbreeding and assortative mating depending on the specific culture.

Most of this book focuses on the four **evolutionary forces**, those mechanisms that lead to a change in allele frequency over time (microevolution). The four evolutionary forces, covered in *much* more detail in later chapters, are

1. **Mutation**. This is a random change in the actual genetic code, including changes in single DNA bases, insertion or deletion of DNA sequences, and other rearrangements of DNA sequences. Mutation provides the ultimate source of all genetic variation, although generally in small amounts in any given generation. The other three evolutionary forces act on mutation, sometimes increasing its frequency and sometimes decreasing its frequency.

2. **Genetic drift**. As noted earlier, reproduction involves probability. Because of change, two parents may not wind up with the expected distribution of genotypes in their offspring. At the level of an entire population, this means that each generation may not have the exact same set of allele frequencies as the previous generation. Allele frequencies can "drift" up and down over time, introducing a random element in microevolution.

3. **Natural selection**. This is Charles Darwin's contribution to evolutionary theory, referring to differential survival and reproduction. Natural selection occurs when there are differences in fitness (the probability of surviving and reproducing) for different genotypes, such that there is a change in allele frequency over time. The example described earlier regarding changes in the color of peppered moths in England is an example of natural selection.

4. **Gene flow**. This is the movement of genes from one population to another. If you move to another population and have a child, your genes have moved as well. Gene flow can introduce new alleles into a population from elsewhere, and can cause populations to be more similar to each other genetically.

The action and interaction of the evolutionary forces are affected by numerous demographic and ecological factors, which are, in turn, affected by human cultural behaviors. Some of the many factors to consider in microevolution include population size, population distribution, the age structure of a population, sex ratios, migration rates, birth rates, disease susceptibility, modes of subsistence, predator–prey relations, and mate choice. Human cultural behavior is connected to each of these factors and others. Although definitions of the term *culture* have been debated by anthropologists for over a century or more, I find that a simple definition applies well in many cases, including the study of human population genetics. Here, **culture** is shared and learned behavior. Culture consists of all of those ideas, customs, and behaviors that an one acquires through the learning process during one's life. Although humans are not the only cultural species (e.g., Whiten et al. 1999), it is clear that we have complex cultures that affect our numbers, distribution, and survival on the planet, and hence affect the genetic structure of our species.

To give only one example, consider population size. The size of a population affects the number of mutations with each generation, and the rapid explosion of our species' numbers in recent historical times has meant that a huge number of potential mutations are generated each generation (Hawks et al. 2007). This rapid and unprecedented population explosion was the product of a number of factors such as the improved agricultural capability of our species and lowered death rates accompanying modernization. Without going into any detail, it is clear that our species' cultural adaptations have altered our population size and carrying capacity, which, in turn, could affect the introduction of new mutations.

Population size is also the critical parameter of genetic drift. As will be described in more detail in Chapter 5, smaller populations experience greater levels of genetic drift. When we consider the limitations on population size of a hunting–gathering lifestyle that all of our ancestors practiced until only 12,000 years or so ago, it is clear that the vast majority of human evolution occurred in a time when genetic drift would have had a major impact on genetic variation both within and between populations. As human populations have altered their way of life through cultural adaptations (agriculture, civilization, industrialization), the level of genetic drift also changed.

Population size is only one example of how culture affects the genetics of human populations. Another (quick) example is adaptation. In evolutionary terms, we generally consider genetic adaptation through natural selection as a population adapts to a specific environment. As environments change, species must also change or become extinct. Many studies of natural selection have looked at how populations adapt to changing environments, such as the changes in coloration of the peppered moth described earlier (e.g., Cook et al. 1999) and studies of finches adapting to climatic and environmental fluctuations (e.g., Weiner 1994). When looking at human populations, we also have to deal with the cultural dimension of adaptation, ranging from the technologies that we use to live in varied climates, to modes of producing food, to our medical responses to disease, among many others. Given the increasingly rapid rate of cultural change in humanity, it is clear that cultural dynamics greatly influence on our genetic variation. Case studies later in the book, ranging from the effect of early agriculture on the spread of hemoglobin mutants to the rapid and independent spread of mutant forms of the lactase gene, will be presented to illustrate this biocultural aspect of anthropology.

C. A Short History of Human Population Genetics

The initial development of the field of population genetics took place well outside the field of anthropology. Much of the core of population genetics can be traced back to the work of three men in the early parts of the twentieth century, Ronald

Fisher (1890–1962), JBS Haldane (1892–1964), and Sewall Wright (1889–1988). Over time, their mathematical formulations were combined with observations from laboratory experiments, field studies, and the fossil record to develop what is often referred to as the *synthetic theory of evolution*, referring to the synthesis of information from a variety of biological and geological fields (Provine 1971).

By the middle of the twentieth century, a number of studies in population genetics had been undertaken on human populations as the number of available genetic markers of the blood increased. Such studies soon realized the rich interplay between cultural, demographic, geographic, historic, and genetic aspects of a population. A classic example is the study of genetic drift in the Parma Valley, Italy by Cavalli-Sforza and colleagues (Cavalli-Sforza 1969; Cavalli-Sforza et al. 2004). At about this time, geneticist Derek Roberts, recognizing the utility of a connection between anthropology and population genetics, coined the term *anthropological genetics*, which has become widely used (including the name of an organization, the American Association of Anthropological Genetics; http://www.anthgen.org/).

Since that time, several key books have been published that directly focus on studies on human population genetics, including *Methods and Theories of Anthropological Genetics* (Crawford and Workman 1973) and three volumes of the series *Current Developments in Anthropological Genetics* (Mielke and Crawford 1980; Crawford and Mielke 1982; Crawford 1984). The late 1960s and early 1970s also lead to two significant textbooks focusing on human population genetics, *Genetics of Human Populations* by Cavalli-Sforza and Bodmer (1971), and *Genetics, Evolution, and Man*, by Bodmer and Cavalli-Sforza (1976). Another influential work that speaks to the status of human population genetics at that time is Morton's (1973) edited volume *Genetic Structure of Populations*. Although it does not deal exclusively with human populations, many of the contributions in the book use data from human populations contributed by the first generation of anthropological geneticists. More recent works include *Anthropological Genetics* (Crawford 2007) and *Human Evolutionary Genetics* (Jobling et al. 2004), both of which offer comprehensive reviews of state-of-the-art methods. Although numerous papers and monographs on human populations have been published in recent decades, I mention the above to give examples of single collections that can be used to get a comprehensive view of the status of the field at different points in its history.

V. A CLOSING THOUGHT

The singular focus on a single species (and our close relatives) often seems strange, given the millions of species on our planet today, although in our own defense, I have to point out that anthropologists do not find this strange at all! Before moving on to the next chapter, I want to reiterate that although my emphasis in this book is on *human* population genetics (I am, after all, an anthropologist), the basic principles of population genetics apply across a wide range of species. It is true that certain methods are more or less applicable to humans than to other species, but the basic mechanisms apply to other species as well.

CHAPTER 2

HARDY–WEINBERG EQUILIBRIUM

An interesting question was raised following a lecture by Reginald Punnett on February 28, 1908. Statistician Udny Yule (1871–1951) asked about the distribution of brachydactyly, an inherited condition where the fingers or toes are very short. Most types of brachydactyly result from a dominant allele, and Yule wondered why, if the trait were dominant, it was not more common in our species. Yule hypothesized that there should be three cases of brachydactyly to every one person with the normal phenotype, which was clearly not the case (Stern 1965). This is an interesting question, and one that I have been asked in my introductory biological anthropology course following my standard review of dominant and recessive alleles.

At first glance, it might seem to make sense that a dominant trait will become more common in the population over time. The reason why it does not, and a fundamental basis of microevolutionary theory, can be understood by looking at what has become known as *Hardy–Weinberg equilibrium* or the *Hardy–Weinberg law*. Following Yule's question, Punnett turned for assistance to his colleague and cricket-playing friend, mathematician Godfrey Hardy (1877–1947). Hardy was apparently interested in pure mathematics, and as Punnett noted, "Knowing that Hardy had not the slightest interest in genetics I put my problem to him as a mathematical one" (Stern 1965:220). Hardy quickly solved the problem mathematically, and published his treatment of Mendelian proportions in populations in 1908, where he commented that "I am reluctant to intrude in a discussion concerning matters of which I have no expert knowledge, and I should have expected the very simple point that I wish to make to have been familiar to biologists" (Hardy 1908:49).

Hardy was too humble, not realizing that his mathematical insight had solved a basic dilemma in genetics. Recognizing the significance of Hardy's insights for the development of genetics, Punnett began referring to the principle as *Hardy's law* (Stern 1965). Later, it was found that a German physician, Wilhelm Weinberg (1862–1937), had arrived independently at the same conclusion and had presented it in a lecture in 1908. In order to recognize both men's independent

Human Population Genetics, First Edition. John H. Relethford.
© 2012 Wiley-Blackwell. Published 2012 by John Wiley & Sons, Inc.

accomplishments, geneticist Curt Stern suggested that the basic principle be henceforth known as the Hardy–Weinberg law (Stern 1943). As you undoubtedly noted from the title of this chapter, Stern's wish has become reality.

I. GENOTYPE AND ALLELE FREQUENCIES

Before getting into the definition, derivation, and application of **Hardy–Weinberg equilibrium**, some basic ideas for measuring genetic variation in a population need to be clarified, specifically the ideas of **genotype frequencies** and **allele frequencies**.

A. Computing Genotype Frequencies

Once again, we use the standard model of a hypothetical locus with two alleles, *A* and *a*. As covered in the previous chapter, when we have two alleles, we will have three genotypes: *AA, Aa*, and *aa*. To make things easy, let us suppose that the two alleles are codominant, so that we can tell the difference between people having these three genotypes. Imagine that we visit a population and test the genotype of 150 people, and obtain the following numbers:

$$AA = 54$$

$$Aa = 72$$

$$aa = 24$$

The first thing we would want to do is to figure out the genotype frequencies, which are the proportions of each genotype. To do this, we simply divide each number by the total number of individuals. Thus, because 54 of 150 people have the *AA* genotype, the genotype frequency is $\frac{54}{150} = 0.36$. We can do this for all three genotypes:

$$AA = \tfrac{54}{150} = 0.36$$

$$Aa = \tfrac{72}{150} = 0.48$$

$$aa = \tfrac{24}{150} = 0.16$$

Although it might seem obvious, it will be important later to see that the sum of the genotype frequencies adds up to 1 ($0.36 + 0.48 + 0.16 = 1.0$). Although we use genotype frequencies in population genetics, note that we could also express these proportions as percentages by multiplying each proportion by 100: $AA = 36\%, Aa = 48\%$, and *aa* $= 16\%$. Note that these add up to 100%.

At this point, it is useful to consider another characteristic of genotype frequencies. Suppose that you were to return to this population and choose a single person at random without knowing that individual's genotype. What is the probability that this person would have the genotype *AA*? The answer is 0.36, which is the genotype frequency, because this number also represents the number of times that a specific event occurs (54 people have genotype *AA*) divided by the total number of events (there are 150 people).

B. Computing Allele Frequencies

You can see that it is easy to estimate out the relative frequency of any genotype; you simply divide the number by the total number of all genotypes. What about the frequency of an allele? What are the number of A alleles and the number of a alleles in the hypothetical population described above?

Two Alleles

The answer to these questions may not be immediately obvious. I have found that many students understand better the concept of allele frequency using an analogy. Imagine there are 150 people each wearing two socks (one on each foot). For reasons unknown to us (you are welcome to make one up), some of the people are wearing two black socks, some are wearing a black sock and a white sock, and some are wearing two white socks. Suppose that you count them up and come up with the following numbers:

> Number of people wearing two black socks = 54
> Number of people wearing one black sock and one white sock = 72
> Number of people wearing two white socks = 24

Now, how many black socks are there in this group of people? If you rush in answering this question, you might come up with 126 people by adding the 54 people, with two black socks and the 72 people with one black sock, but that is actually the number of people having at least one black sock, which is not what I asked. Instead, you have to count the actual number of black socks, keeping in mind that some people are wearing two and some people are wearing only one. In this example, we have 54 people, each with two black socks, and 72 people with only one black sock. When we add these up, we get a total of

$$54(2) + 72(1) = 108 + 72 = 180$$

We can do the same thing for the number of white socks. We have 72 people with one white sock and 24 people with two white socks, giving a total of

$$72(1) + 24(2) = 72 + 48 = 120$$

We have 180 black socks and 120 white socks, for a total of 300 socks, which makes sense because we have a total of 150 people, each with 2 socks [$= 150(2) = 300$]. We could now figure out the relative frequency of black socks by dividing the number of black socks by the total number of socks, which is $\frac{180}{300} = 0.6$. Likewise, the relative frequency of white socks is $\frac{120}{300} = 0.4$.

You probably noticed that the numbers I used in the genotype frequency example and the sock example are the same. This was done on purpose to enable you to extend the sock analogy to the concept of allele frequencies. The computation is the same, except that instead of counting socks you are counting alleles. Let us return to the original question. Given the genotype numbers $AA = 54, Aa = 72$, and $aa = 24$, what are the relative frequencies of the A and a alleles? Start by counting the number of A alleles, remembering to count the A allele twice for the AA genotype and once for the Aa genotype. There are $54(2) + 72(1) = 180\ A$ alleles. We repeat this procedure for the number of a alleles, getting $72(1) + 24(2) = 120\ a$ alleles. Thus, this hypothetical population has $180\ A$ alleles and $120\ a$ alleles for a total of 300 alleles (which works out since there are 150 people, each with two alleles). Therefore, the relative frequency of the A allele is $\frac{180}{300} = 0.6$, and the relative frequency of the a allele is $\frac{120}{300} = 0.4$. A simple way to keep all of these calculations straight is to make a table that lists the number of A alleles and the number of a alleles for each of the three genotypes, as shown in Example 2.1.

EXAMPLE 2.1   How to Compute Allele Frequencies.   In this example, we use the hypothetical example from the text of a single locus with two alleles, A and a. The data are

$$\text{Number of people with genotype } AA = 54$$

$$\text{Number of people with genotype } Aa = 72$$

$$\text{Number of people with genotype } aa = 24$$

In the following table, we list the number of A and a alleles and the total number of all alleles for each genotype, and then sum each column:

Genotype	Number of People	Number of A Alleles	Number of a Alleles	Number of all Alleles
AA	54	108	0	108
Aa	72	72	72	144
aa	24	0	48	48
Total	150	180	120	300

There are a total of 180 A alleles and 120 a alleles in this population, for a total of 300 alleles (twice the number of people sampled, because each person has two alleles).

$$\text{Relative frequency of } A \text{ allele } = \tfrac{180}{300} = 0.6$$

$$\text{Relative frequency of } a \text{ allele } = \tfrac{120}{300} = 0.4$$

As a check, the allele frequencies should add up to 1.0, which they do ($0.6 + 0.4 = 1.0$).

Because we use allele frequencies extensively in population genetics equations, we use symbols to refer to the different allele frequencies. Although you can feel free to create any symbol you wish, the conventional format is to use the symbol p to refer to one allele and the symbol q to refer to the other allele. In Example 2.1, p is the relative frequency of the A allele and q is the relative frequency of the a allele. We will use this notation throughout the rest of the book.

Note that in the case of two alleles, there are only two allele frequencies, and these numbers must add up to 1.0. In Example 2.1, this is seen because

$0.6 + 0.4 = 1.0$. This property allows us a convenient check on our math; if the two allele frequencies do not add up to 1.0 (or very close, in the case of irrational numbers and roundoff), then an error was made in computing the allele frequency. Beware—I have seen this error on homework assignments, so you should always take time to check your answers thoroughly.

Mathematically, we can express the sum of the allele frequencies using the following simple formula:

$$p + q = 1 \tag{2.1}$$

Note that this relationship allows us to predict one allele frequency if we know the other, because, using some simple algebra, we see that

$$p = 1 - q$$
$$q = 1 - p \tag{2.2}$$

These relationships may be intuitively obvious, but as you will see, they allow us to handle much of the mathematics of population genetics in an elegant manner.

As another example of computing allele frequencies, I will use a real-world case. In the 1960s, anthropologist Jonathan Friedlaender collected data on a number of genetic markers of the blood in 18 villages on the island of Bougainville in Melanesia. One of the markers collected was the haptoglobin locus, a gene located on chromosome 16. This locus has two alleles, *HPA*1* and *HPA*2*. In the village of Nupatoro, Friedlaender (1975) collected blood samples from 111 villagers, and found the following genotype numbers:

$$HPA^*1\text{–}HPA^*1 = 22$$
$$HPA^*1\text{–}HPA^*2 = 65$$
$$HPA^*2\text{–}HPA^*2 = 24$$

The genotype frequencies are obtained by dividing each number by the total number:

$$HPA^*1\text{–}HPA^*1 = \tfrac{22}{111} = 0.198$$
$$HPA^*1\text{–}HPA^*2 = \tfrac{65}{111} = 0.586$$
$$HPA^*2\text{–}HPA^*2 = \tfrac{24}{111} = 0.216$$

We find that there are 109 *HPA*1* alleles and 113 *HPA*2* alleles, for a total of 222 alleles (twice the number of people sampled). This gives allele frequencies for *HPA*1* of $\tfrac{109}{222} = 0.491$ and for *HPA*2* of $\tfrac{113}{222} = 0.509$. The full computation is provided in Example 2.2 for review.

EXAMPLE 2.2 How to Compute Allele Frequencies. This example is based on an actual study of human population genetics. Data on a number of genetic markers were collected by Friedlaender (1975) on Bougainville Island in Melanesia. One of these markers was the red blood cell protein haptoglobin protein, a locus with two alleles: HPA^*1 and HPA^*2. Genotype numbers and number of alleles are shown below:

Genotype	Number of People	Number of HPA^*1 Alleles	Number of HPA^*2 Alleles	Number of All Alleles
$HPA^*1–HPA^*1$	22	44	0	44
$HPA^*1–HPA^*2$	65	65	65	130
$HPA^*2–HPA^*2$	24	0	48	48
Total	111	109	113	222

$$\text{Relative frequency of } HPA^*1 \text{ allele} = p = \tfrac{109}{222} = 0.491$$

$$\text{Relative frequency of } HPA^*2 \text{ allele} = q = \tfrac{113}{222} = 0.509$$

More than Two Alleles

Much of population genetics theory builds on a simple model of a single locus with two alleles. In the real world, however, there are many loci (particularly DNA markers) with more than two alleles. How can we compute allele frequencies in such cases? For loci where all the alleles are codominant, allowing identification of each genotype, the answer is simple—we count the number of alleles in the same manner. Example 2.3 shows computations for a locus with three alleles. When we have three alleles, we typically label the allele frequencies as p, q, and r (when there are more than three alleles, it is common to simply use the letter p with a subscript to refer to different alleles).

EXAMPLE 2.3   How to Compute Allele Frequencies for a Locus with Three Alleles.   Salzano et al. (1985) collected data from 216 Pacaás Novos Indians of Brazil on the group-specific component locus (GC, also known as the *vitamin D binding protein*) on chromosome 4, which codes for a protein. A number of early studies indentified two alleles, GC^*1 and GC^*2, but later work identified two subtypes of the GC^*1 allele, known as GC^*1F and GC^*1S. Salzano et al. collected data on the six genotypes associated with the three alleles GC^*1F, GC^*1S, and GC^*2:

Genotype	Number of People	GC^*1F Alleles	Number of GC^*1S Alleles	Number of GC^*2 Alleles	Number of All Alleles
$GC^*1F–GC^*1F$	43	86	0	0	86
$GC^*1F–GC^*1S$	75	75	75	0	150
$GC^*1S–GC^*1S$	32	0	64	0	64
$GC^*2–GC^*1F$	34	34	0	34	68
$GC^*2–GC^*1S$	21	0	21	21	42
$GC^*2–GC^*2$	11	0	0	22	22
Total	216	195	160	77	432

By summing up the columns, we see that there are 195 *GC*1F* alleles, 160 *GC*1S* alleles, and 77 *GC*2* alleles for a total of 432 alleles.

$$\text{Relative frequency of } GC*1F \text{ allele } = p = \tfrac{195}{432} = 0.451$$

$$\text{Relative frequency of } GC*1S \text{ allele } = q = \tfrac{160}{432} = 0.370$$

$$\text{Relative frequency of } GC*2 \text{ allele } = r = \tfrac{77}{432} = 0.178$$

Note that the sum of the three allele frequencies does not add up exactly to 1.0 ($0.451 + 0.370 + 0.178 = 0.999$). This is because of roundoff error; if you do all the computations of the allele frequencies to four decimal places, you will see that they do add up to 1.0 ($0.4514 + 0.3704 + 0.1782 = 1.0$).

An Alternative Method

The method used so far is often called the *allele counting method* because it relies on counting the number of different alleles over all genotypes. Another method of computing allele frequencies is based on the genotype frequencies. Returning to the first example used in this chapter, we have a locus with two alleles, *A* and *a*, in a hypothetical population of 150 people with the following genotypes: 54 *AA*, 72 *Aa*, and 24 *aa*. When we counted the alleles, we found 180 *A* alleles and 120 *a* alleles, giving allele frequencies of $p = \tfrac{180}{300} = 0.6$ and $q = \tfrac{120}{300} = 0.4$.

Here is another way to calculate the allele frequencies. First, compute the genotype frequencies as before by dividing the number in each genotype by the total number of genotypes ($= 150$). We did these calculations earlier in this chapter and obtained:

$$f_{AA} = \tfrac{54}{150} = 0.36$$

$$f_{Aa} = \tfrac{72}{150} = 0.48$$

$$f_{aa} = \tfrac{24}{150} = 0.16$$

Note that I have assigned the symbol *f* to refer to *genotype frequency* and used subscripts to refer to the different genotypes. Thus f_{AA} refers to the frequency of the *AA* genotype, f_{Aa} refers to the frequency of genotype *Aa*, and f_{aa} refers to the frequency of genotype *aa*. We can now compute the frequency of the *A* allele by adding the frequency of the *AA* genotype to *half* the frequency of the *Aa* genotype. Likewise, we can compute the frequency of the *a* allele by adding the frequency of *half* the *Aa* genotype to the frequency of the *aa* genotype. This is easier to express mathematically than in words (which is *why* we use symbols):

$$p = f_{AA} + \frac{f_{Aa}}{2}$$

$$q = \frac{f_{Aa}}{2} + f_{aa} \tag{2.3}$$

When we plug in the actual genotype frequencies in our example, we get the same allele frequencies as with the allele counting method:

$$p = 0.36 + \frac{0.48}{2} = 0.36 + 0.24 = 0.6$$

$$q = \frac{0.48}{2} + 0.16 = 0.24 + 0.16 = 0.4$$

Note that this method may not give the exact same results as the allele counting method if you did not compute the genotype frequencies using enough decimal places to avoid roundoff problems. My rationale for introducing this alternative method will be clear a little bit later.

Can you show *why* equation (2.3) works mathematically? The proof is shown in Appendix 2.1 at the end of this chapter.

II. WHAT IS HARDY–WEINBERG EQUILIBRIUM?

What exactly is the Hardy–Weinberg law, and why is it so important? It is difficult to explain Hardy–Weinberg equilibrium clearly and comprehensively in a short definition, which is why I have devoted an entire chapter to it. At the most basic level, Hardy–Weinberg is an mathematical expression describing the expected genotype frequencies in a new generation. In the previous chapter, I reviewed how Punnett squares are used to compute the expected genotype distribution in offspring given the parent's genotypes. Hardy–Weinberg provides a way of extending this idea to an entire population. Imagine that we have a locus with two alleles, A and a, where p is the frequency of the A allele and q is the frequency of the a allele. Both Hardy and Weinberg independently showed that the expected genotype frequencies in the next generation are

$$\text{Expected frequency of genotype } AA = p^2$$

$$\text{Expected frequency of genotype } Aa = 2pq$$

$$\text{Expected frequency of genotype } aa = q^2 \tag{2.4}$$

These equations are often familiar to the introductory student in biological anthropology and biology classes. Following the initial math fear experienced by some students, just about everyone soon learns these simple equations. For example, if we know that p = 0.7 and q = 0.3 in the parental gene pool, we can quickly figure out the expected distribution of genotypes in the next generation as

$$\text{Expected frequency of genotype } AA = p^2 = (0.7)^2 = 0.49$$

$$\text{Expected frequency of genotype } Aa = 2pq = 2(0.7)(0.3) = 0.42$$

$$\text{Expected frequency of genotype } aa = q^2 = (0.3)^2 = 0.09$$

The utility of predicting outcomes in the next generation can be illustrated with a more specific type of question. For example, let us imagine that the frequency of a harmful recessive allele is 0.01 and we want to know how many children will be born having two recessives, and hence a genetic disease. In this case, Hardy–Weinberg allows a quick answer: $q^2 = (0.01)^2 = 0.0001$, which is 1 in 10,000 offspring. Likewise, we could compute the expected proportion of heterozygotes that would not get the disease but would be carriers: $2pq = 2(0.99)(0.01) = 0.0198$, which is 198 in 10,000 offspring.

Using the Hardy–Weinberg equations is straightforward. It is a second aspect of the Hardy–Weinberg law that causes more confusion. Both Hardy and Weinberg independently showed that, under certain conditions, the genotype and allele frequencies would remain constant from one generation to the next. In other words, Hardy–Weinberg equilibrium states that nothing changes (the very definition of *equilibrium*). Another way of saying that nothing changes is saying that there is no evolution! At this point in a typical lecture, everyone becomes rightfully perplexed. It is obvious from lab experiments, field studies, and the fossil record that organisms evolve all the time, which makes it difficult to understand why valuable lecture time (and textbook space) is being taken up by something that is clearly incorrect. The answer, which we then give in lecture, is that Hardy–Weinberg equilibrium gives us a *baseline*; we start with the case where nothing happens to show how something could happen! Although true, the underlying nature and utility of Hardy–Weinberg equilibrium often gets lost in the introductory course. Weiss and Kurland (2007:204) sum the confusing nature of Hardy–Weinberg equilibrium quite nicely, noting, "As one student griped, 'Let me get this straight. When nothing happens . . . nothing changes? Duh.'"

I think that a major problem with understanding Hardy–Weinberg equilibrium is that it is difficult to reduce the concept and its implications into a simple but understandable definition. The idea instead takes more time and development, which is the goal of the remainder of this chapter.

III.   THE MATHEMATICS OF HARDY–WEINBERG EQUILIBRIUM

Given allele frequencies p and q, equation (2.4) allows us to predict the expected genotype frequencies in the next generation. Where did equation (2.4) come from? There are several ways to answer this, and Hardy and Weinberg each used a somewhat different method to answer this question.

A.   A Simple Proof

Here, I will start by demonstrating Hardy–Weinberg proportions [equation (2.4)] using a simple model of probability that uses an analogy for allele frequencies. Picture a cup filled with 100 small plastic beads, of which 60 are red and 40 are blue. We can say that the relative frequency of red beads is $\frac{60}{100} = 0.6$ and the relative frequency of blue beads is $\frac{40}{100} = 0.4$. Imagine that all of these beads are mixed together and you pull one bead from the cup at random. What is the probability of getting a red bead? It is 0.6 (and the probability of getting a blue bead is 0.4). Now, imagine something a little more complicated. Pull out a bead, replace it in the cup, mix then beads together, and then pull out another bead. What is the probability of getting a red bead *both* times? We answer this using the AND rule of probability from the previous chapter. The probability of getting two red beads is the probability of getting a red bead (0.6) multiplied by the probability of getting a red bead (0.6), which is $0.6 \times 0.6 = 0.36$. We can do the same type of computation to answer the question of repeating the same experiment and getting two *blue* beads. Here, the probability would be $0.4 \times 0.4 = 0.16$.

What about getting one red bead and one blue bead? This is a little more complicated because there are two ways of getting a red bead and a blue bead. The first way is to get a red bead on the first try and a blue bead on the second try, and the second way is the reverse—getting a blue bead on the first try and a red bead on the second try. We can solve this problem by breaking it down into several steps. First, what is the probability of getting a red bead on the first try and a blue bead on the second try? We use the AND rule to figure this out and multiply the probability of getting a red bead (0.6) by the probability of getting a blue bead (0.4), which is $0.6 \times 0.4 = 0.24$. The second step is to determine the probability of the reverse happening, where we get a blue bead on the first try and a red bead on the second. Using the AND rule, we get $0.4 \times 0.6 = 0.24$. Thus, we have a probability of 0.24 of getting a red bead followed by a blue bead, AND we have a probability of 0.24 of getting a blue bead followed by a red bead. However, my initial question was only about the overall probability of getting one red bead and one blue bead, and *the order does not matter*. What is the probability of getting a red bead and then a blue bead OR a blue bead and then a red bead? Here, we use the OR rule and add the probabilities. The probability of getting one red bead and one blue bead, regardless of the order, is therefore $0.24 + 0.24 = 0.48$.

We can now summarize the probability distribution of taking a bead from the cup at random, replacing it, and then taking a bead again at random:

Probability of getting two red beads $= 0.36$

Probability of getting one red bead and one blue bead $= 0.48$

Probability of getting two blue beads $= 0.16$

Note that the sum of these probabilities is 1 $(0.36 + 0.48 + 0.16)$, because we have listed all possible outcomes.

Going back to genetics, we can use the same principle. Instead of an experiment where we take a bead at random from a cup of beads, picture the process of reproduction of an allele that gets into the next generation as equivalent to taking an allele at random from a gene pool. Given two alleles, A and a, the possible genotypes in the next generation are AA, Aa, and aa. How do you get the AA genotype? The answer is when each parent contributes an A allele. To put this another way, what is the probability of a child getting an A allele from one parent and an A allele from the other parent? Given that p represents the frequency of the A allele, and is the probability of a given individual inheriting an A allele, the expected frequency for genotype AA in the next generation is p times $p, = p^2$.

How about the genotype Aa? This can happen in one of two ways. The father can contribute an A allele and the mother can contribute an a allele, or the reverse can occur, where the father contributes an a allele and the mother contributes an A allele. The probability of the first situation occurring is p times $q = pq$. The probability of the second situation occurring is q times $p = qp$, which by the commutative law of algebra is the same as pq. Thus, the probability of getting an A allele from the father and an a allele from the mother *or* the reverse is $pq + pq = 2pq$.

The final genotype to consider is aa. The probability of obtaining this genotype is computed as the probability of both parents contributing an a allele, which is q times $q = q^2$.

These proportions ($AA = p^2, Aa = 2pq, aa = q^2$) are the expected genotype frequencies [shown in equation (2.4)] expected under Hardy–Weinberg equilibrium. Table 2.1 summarizes the logic used here to demonstrate Hardy–Weinberg frequencies. Another method of deriving Hardy–Weinberg frequencies is to consider the genotypes for all possible offspring resulting from all possible random mating of parental genotypes (e.g., $AA \times AA, AA \times Aa$). This derivation can be found in most advanced population genetics texts (e.g., Hartl and Clark 2007).

Figure 2.1 shows the nonlinear relationship between allele frequencies and genotypes frequencies. As p increases, the proportion of homozygous genotype AA increases geometrically, and the frequency of homozygous genotype aa decreases geometrically. Note that the frequency of heterozygotes (Aa) increases

TABLE 2.1   Using Simple Probability to Derive Hardy–Weinberg Proportions[a]

Genotype of Child	Allele from Father		Allele from Mother	Probability
AA	A		A	$p \times p = p^2$
Aa	A	or	a	$p \times q = pq$
	a		A	$q \times p = qp = pq$ $\Big\}= pq + pq = 2pq$
aa	a		a	$q \times q = q^2$

[a]The data presented in this table are based on a single locus with two alleles, A and a, where p is the frequency of the A allele in the parental population and q is the frequency of the a allele in the parental population.

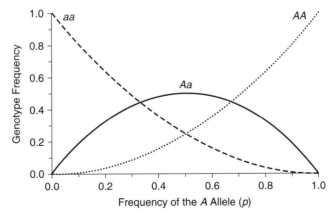

FIGURE 2.1 Genotype frequencies under Hardy–Weinberg equilibrium for a single locus with two alleles (A and a). The frequency of the A allele is p, and the frequency of the a allele is $q = 1 - p$. The genotype frequencies are $AA = p^2, Aa = 2pq$, and $aa = q^2$.

as p increases to reach a maximum at $p = 0.5$, after which the frequency of heterozygotes decreases.

For Example 2.1, where $p = 0.6$ (frequency of the A allele) and $q = 0.4$ (frequency of the a allele), the expected genotype frequencies in the next generation are $AA = p^2 = 0.36$, $Aa = 2pq = 0.48$, and $aa = q^2 = 0.16$. One last thing. Just as the allele frequencies must add up to 1.0 ($p + q = 1$), so must the genotype frequencies:

$$p^2 + 2pq + q^2 = 1 \qquad (2.5)$$

This may be intuitive, but we could also demonstrate this by noting that because $q = 1 - p$, we can substitute this into equation (2.5). After doing all the algebra, the answer is 1.

B. What Does Equilibrium Mean?

So far, we have started with a parental population with allele frequencies p and q that we use to compute the expected genotype frequencies in the next generation. What happens then? The next thing we need to know is the allele frequencies in the offspring generation. Here we use the symbols p' and q' to refer to the allele frequencies in the offspring, using the prime symbol (′) to differentiate these from the parental allele frequencies (p and q). We can figure out the allele frequencies of the offspring from their genotype frequencies using equation (2.3) with the expected Hardy–Weinberg frequencies from equation (2.4); that is, to compute p' we take the expected frequency of the AA genotype and add half the expected frequency of the Aa genotype:

$$p' = f_{AA} + \frac{f_{Aa}}{2} = p^2 + \frac{2pq}{2}$$

When we factor out the p in this equation and let the number 2 in the numerator and denominator cancel out, we get $p' = p(p + q)$, and because $p + q = 1$ [equation (2.1)], we obtain $p' = p$. We could do the same thing to compute the frequency of the a allele in the next generation and would get $q' = q$. In other words, *there has been no change in the allele frequency*. This is the *equilibrium* part of Hardy–Weinberg equilibrium–equilibrium means no change.

We could continue the process even farther by using the allele frequencies of the offspring to compute the expected genotype frequencies of *their* offspring. Because the allele frequencies of the offspring are the same as those of the parents, the expected genotype frequencies would also be the same as those of the parents: $p^2, 2pq$, and q^2. This process would then continue to the next generation, also resulting in allele frequencies of p and q and genotype frequencies of $p^2, 2pq$, and q^2.

An important thing to remember is that even if the population is not initially at Hardy–Weinberg equilibrium, it will reach an equilibrium state within a single generation. To show this, take the case where we start with genotype frequencies of $AA = 0.33$, $Aa = 0.54$, and $aa = 0.13$. Using equation (2.3), we obtain allele frequencies of $p = 0.6$ and $q = 0.4$. The expected genotype frequencies are $p^2 = 0.36$, $Aa = 2pq = 0.48$, and $aa = q^2 = 0.16$, which are different from what we started with. This means that the initial population is not at Hardy–Weinberg equilibrium. We then use equations (2.3) and (2.4) to compute the allele and genotype frequencies for the next generation. From this point on, the expected genotype frequencies will always be $AA = 0.36, Aa = 0.48$, and $aa = 0.16$, and the allele frequencies will always be $p = 0.6$ and $q = 0.4$. (Try it!) In other words, if a population does not begin in Hardy–Weinberg equilibrium, it will reach Hardy–Weinberg equilibrium in a single generation.

If a population is in Hardy–Weinberg equilibrium, allele and genotype frequencies will remain unchanged over time. The equilibrium of allele frequencies is particularly interesting, because given that evolution can be defined as a change in allele frequencies, Hardy–Weinberg equilibrium predicts that there is no change, and therefore that evolution does *not* occur! However, we see evidence of changing allele frequencies (microevolution) all the time in laboratory and field studies, which means that Hardy–Weinberg is clearly incorrect across the boards. Why, then, do we spend so much time (and an entire chapter) looking at Hardy–Weinberg equilibrium? Of what possible use is it?

The main use of Hardy–Weinberg equilibrium is that it provides a baseline for explaining change. Hardy–Weinberg gives us a prediction of what will happen *given certain assumptions*. Returning to the question raised by Yule that was discussed at the start of this chapter, Hardy–Weinberg equilibrium demonstrates that, given certain assumptions, the proportion of people with brachydactyly will *not* change even though the allele for brachydactyly is dominant. As long as the assumptions of Hardy–Weinberg equilibrium are met, there will be no change in genotype or allele frequencies. Therefore, when we *do* see change over time, we know that one or more of the underlying assumptions of Hardy–Weinberg equilibrium are not true. As such, we can use Hardy–Weinberg to model what-if scenarios that provide us with clues about how evolution works. All of the basic concepts of population genetics discussed in this book start with the assumption of Hardy–Weinberg equilibrium and then ask what will happen under a different set of conditions. The equilibrium model is a powerful tool for developing evolutionary models.

C. Assumptions of Hardy–Weinberg Equilibrium

Anytime we use a mathematical model of reality, we will have to make a number of assumptions, and Hardy–Weinberg equilibrium is no exception. In this case, we assume that we are dealing with a sexually reproducing species (two sexes). We also assume that each generation is discrete and does not overlap with any other generation; in other words, no one from one generation mates with anyone from another generation.

A key assumption of Hardy–Weinberg equilibrium is random mating. This means that the probability of getting a given genotype is purely a function of the allele frequencies and the application of the laws of probability described earlier. In reality, the assumption of random mating can be violated in either of two ways. One exception to random mating is inbreeding, the mating between closely related individuals. Although everyone is inbred to some extent (see Chapter 3), we are generally concerned with mating between close relatives, such as between siblings, aunts and nephews, uncles and nieces, and close cousins. Under inbreeding, the probability of someone being homozygous increases because of the closer than random relationship of the parents. We expect a certain proportion of homozygotes by chance under Hardy–Weinberg equilibrium ($AA = p^2$ and $aa = q^2$). Under inbreeding, these proportions will be larger than expected by chance, as will be shown in the next chapter.

Another way in which random mating is violated is when there is a preference for choosing a mate according to someone's genotype. In our hypothetical example, this means that someone with genotype AA prefers a mate that has the same genotype, such that $AA \times AA$ mating will occur more frequently than expected by chance. Hardy–Weinberg equilibrium assumes that this is not the case, and that the probability of a given mating is purely a function of probability. There is no preference for choosing a mate on the basis of their genotype. This is true for many (but not all) traits. (Do you think that people are concerned with their possible mate's ABO blood type?) Sometimes people choose mates according to on their phenotypic similarity (and, hence, some underlying genetic similarity), which is known as *positive assortative mating*. Some examples in humans include mate choice based on similar skin color, weight, and height. As with inbreeding, positive assortative mating will increase the proportions of homozygotes.

Deviations from random mating will lead to genotype frequencies that are different from those expected under Hardy–Weinberg equilibrium. However, these deviations will *not* change the actual allele frequencies (the reason for this will be demonstrated in Chapter 3).

Violation of the other assumptions of Hardy–Weinberg equilibrium *will* cause a change in allele frequencies. One of these assumptions is that there is no mutation. In our simple example, we have a locus with two alleles, A and a, and Hardy–Weinberg equilibrium assumes that these two alleles will be passed on to the next generation, but that none of the A alleles will mutate into a, none of the a alleles will mutate into A, and neither allele will mutate to a new allele. If this assumption is violated, then there will be a change, albeit small, in the allele frequencies. Suppose, for example, that a population has 60 A alleles and 40 a alleles, giving frequencies of 0.6 and 0.4. Imagine that one of the A alleles mutates into a new form, which we will call B. We now have 59 A alleles, 40 a alleles, and 1 B allele. The frequency of the A allele is now $\frac{59}{100} = 0.59$. Hardy–Weinberg equilibrium assumes that mutation does not happen, but in the real world, there will always be a small fraction of alleles that mutate and are passed on to the next generation. Although small, this is a change in allele frequency. Details on the changes in allele frequency due to mutation are covered in Chapter 4.

Another assumption of Hardy–Weinberg equilibrium is that there are no chance deviations in allele frequency due to sampling error. This idea, which was described briefly in the last chapter, is an extension of probability theory that shows that a sample may not represent the statistical universe from which it was taken. In population genetic terms, genetic drift is the change in allele frequencies that occurs because of sampling error such that, by chance, the allele frequencies in any generation are not likely to be the same as in the previous generation. Hardy–Weinberg equilibrium assumes that there is no genetic drift. As will be shown in more detail in Chapter 5, the expected effect of genetic drift depends on population size—the smaller the population, the greater the likely effect of genetic drift. Hardy–Weinberg equilibrium assumes that there is absolutely no genetic drift, and in essence assumes that the population is of infinite size. Although this is clearly not possible, in practical terms this translates to the population being very large, such that there is no perceptible impact of genetic drift.

Hardy–Weinberg equilibrium also assumes that the chance of survival and reproduction does not depend on someone's genotype; in other words, someone with genotype *AA* is not more or less likely to survive and reproduce than someone with genotype *Aa* or *aa*. Hardy–Weinberg equilibrium assumes that there is no natural selection. In reality, there are times when one genotype might have a higher probability of surviving and reproducing than other genotypes. If one genotype is represented in the next generation more frequently than expected under Hardy–Weinberg equilibrium, then the allele frequencies will change. For example, imagine that individuals with genotype *AA* are more likely to survive and reproduce than the other genotypes. This selection will result in more individuals with genotype *AA* than expected, and therefore more *A* alleles. Thus, the frequency of *A* will increase, while the frequency of *a* will decrease. Chapter 6 presents the outcomes of various different models of natural selection, showing examples where an allele frequency will increase, decrease, or reach a balance. Chapter 7 builds on this theoretical foundation by discussing a number of case studies of natural selection in human populations.

The last assumption of Hardy–Weinberg equilibrium discussed here is the assumption that the population is closed and there are no migrants into the population from another population. This assumption means that there is no gene flow entering the population. What happens genetically in one population does not affect what happens in any other population. The effect of gene flow on populations (which has been major in human evolution) is discussed in more detail in Chapter 8.

Note that last four assumptions (no mutation, no selection, no genetic drift, and no gene flow) correspond to the four evolutionary forces introduced in the previous chapter. All of these involve a change in allele frequencies. We use Hardy–Weinberg equilibrium to establish the baseline, and then change one of more of these assumptions to predict levels and direction of microevolutionary change. We also apply our knowledge of these assumptions to compare observed reality with theoretical expectation to gain insight into how evolution works.

IV. USING HARDY–WEINBERG EQUILIBRIUM

The major use of Hardy–Weinberg equilibrium in this text is the establishment of a baseline condition—no changes in genotype or allele frequencies—that allows us to model what will happen when one or more of the underlying assumptions of Hardy–Weinberg equilibrium are not met. There are also some other uses of Hardy–Weinberg equilibrium.

A. Detecting Deviations from Hardy–Weinberg Equilibrium

In principle, we can compare observed and expected genotype numbers to explicitly test the hypothesis of equilibrium. Imagine that you visit a population of 80 people and find the following genotype numbers for the hypothetical locus that we have been using:

$$AA = 40$$

$$Aa = 32$$

$$aa = 8$$

These are the observed genotype numbers. We can now compare these numbers with the numbers expected under Hardy–Weinberg equilibrium. We do this by computing the allele frequencies and then use Hardy–Weinberg equilibrium to compute the expected genotype numbers.

In this example, we use the allele counting method and find that there are 112 A alleles and 48 a alleles, for a total of 160 alleles. This gives allele frequencies of $p = \frac{112}{160} = 0.7$ and $q = \frac{48}{160} = 0.3$. We now use Hardy–Weinberg equilibrium to predict the genotype frequencies: $AA = p^2 = 0.49, Aa = 2pq = 0.42$, and $aa = q^2 = 0.09$. Finally, we multiply these expected genotype *frequencies* by the total number of people ($= 80$) to get the expected genotype *numbers*: $AA = 0.49(80) = 39.2, Aa = 0.42(80) = 33.6$, and $aa = 0.09(80) = 7.2$. Our comparison of observed and expected numbers is summarized below:

Genotype	Observed Number	Expected Number
AA	40	39.2
Aa	32	33.6
aa	8	7.2
Total	80	80

If the population is at Hardy–Weinberg equilibrium, then the observed genotype numbers will be the same as the expected genotype numbers. Rounding off (to deal with the mathematical possibility, but biological impossibility, of a fractional person), we expect to see 39 AA, 34 $A\,a$, and 7 aa individuals. After rounding off, we see one more person with genotype AA than expected, two more with genotype Aa than expected, and one less with genotype aa than expected.

Is the population at Hardy–Weinberg equilibrium? Although the observed and expected numbers are not exactly the same, they are close. The issue of Hardy–Weinberg equilibrium boils down to a statistical question—is the difference between observed and expected genotype numbers statistically significant? How likely would we be to sample 80 individuals and get these small deviations due to chance? This is similar to many questions in statistics. For example, if the average height of male students in a large lecture class were 5 ft 9 in., then how likely would it be for a sample of, say, 10 students to have this exact same average height?

Consider another example with observed genotype numbers in a population of 130 individuals of $AA = 90, Aa = 28$, and $aa = 12$. Here, we obtain allele frequencies of $p = 0.8$ and $q = 0.2$, and the following comparisons:

Genotype	Observed Number	Expected Number
AA	90	83.2
Aa	28	41.6
aa	12	5.2
Total	130	130

Here, we get differences that are much larger than in the first example. The question, however, is the same—how likely is it to get differences this large by chance? If it is not statistically likely, then we can say that the population is *not* at Hardy–Weinberg equilibrium. In statistical terms, we would use a **chi-square test** to answer the question. Details of this test are presented in Appendix 2.2. As shown there, it turns out that the first example presented above is not significantly different from Hardy–Weinberg equilibrium, but the second one is.

Statistical tests for Hardy–Weinberg equilibrium are actually somewhat limited. Most actual applications of this test show Hardy–Weinberg equilibrium for most loci. As pointed out by Weiss and Kurland (2007), this does not mean that evolution is not occurring, but instead that the typical levels of allele frequency changes that occur in a single generation are unlikely to be detected in typical samples. Hardy–Weinberg equilibrium is not always that useful for detecting microevolution, and its greatest value is in development of models explaining microevolution, as will be outlined throughout this text.

B. Hardy–Weinberg Equilibrium and Dominant Alleles

Hardy–Weinberg equilibrium also has a practical application for computing allele frequencies. The allele counting method for computing allele frequencies works only when the alleles are codominant. If one allele is dominant and the other is recessive, then the method cannot work. To see this, consider PTC tasting, a trait in humans that involves the ability to taste the chemical phenylthiocarbamide. Some people can taste this bitter chemical and others cannot. The genetics of PTC tasting is approximated by a model with two alleles, a dominant *tasting* allele (T) and a recessive *nontasting* allele (t). Using this model, the genotypes and phenotypes are

Genotype	Phenotype
TT	Taster
Tt	Taster
tt	Nontaster

There is no way to count the alleles in this case. If someone is a nontaster, we know that person's genotype is tt and that we would count two t alleles. However, if someone is a taster, all we know is that this individual could have either genotype

TT or genotype *Tt*, and we therefore do not know whether to count one or two *T* alleles.

We can solve this problem by assuming that this locus is at Hardy–Weinberg equilibrium in the population we are studying. If we let *p* be the frequency of the *T* allele and *q* be the frequency of the *t* allele, and then assume Hardy–Weinberg equilibrium, we can write the phenotype frequencies as follows:

$$\text{Frequency of tasters} = \text{frequency of } TT + \text{frequency of } Tt = p^2 + 2pq$$

$$\text{Frequency of nontasters} = \text{frequency of } tt = q^2$$

Because the frequency of nontasters is q^2, we can compute the frequency of q by taking the square root of the frequency of nontasters. Once we have q, we can compute $p = 1 - q$ [from equation (2.2)].

Here is an example. Giles et al. (1968) collected data on PTC tasting for 1374 people in the town of Ticul, Yucatán, Mexico, finding 1257 tasters and 117 non-tasters. The frequency of non-tasters is $\frac{117}{1374} = 0.085$. We take the square root of this frequency to get the frequency of the *t* allele:

$$q = \sqrt{q^2} = \sqrt{0.085} = 0.292$$

The frequency of the *T* allele is therefore

$$p = 1 - q = 1 - 0.292 = 0.708$$

We can compute allele frequencies when there is dominance and as long as we assume Hardy–Weinberg equilibrium. Of course, if we make this assumption, we cannot then test for equilibrium, and the observed and expected numbers will always be the same! Methods that are more complex exist for situations where there are more than two alleles and one of them is dominant.

V. EXTENSIONS OF HARDY–WEINBERG EQUILIBRIUM

There are several extensions of Hardy-Weinberg equilibrium. Three of these are introduced briefly here.

A. Linkage Disequilibrium

So far, we have been dealing with a single locus. There are many useful methods in population genetics that look at the association of two loci. Imagine that we have two loci, each with two alleles, where the first locus has alleles *A* and *a*, and the second locus has alleles *B* and *b*. To keep track of the allele frequencies, we will use *p* and *q* with subscripts, where subscript 1 refers to the first locus and subscript 2 refers to the second locus. The allele frequencies for these two loci are thus labeled as follows:

$$p_1 = \text{frequency of allele } A$$

$$q_1 = \text{frequency of allele } a$$

$$p_2 = \text{frequency of allele } B$$

$$q_2 = \text{frequency of allele } b$$

Different combinations (haplotypes) are possible. A given **gamete** (sex cell) could have one of four combinations for the two loci: *AB, Ab, aB,* and *ab*. We use the symbol *X* to represent the frequency of a given gamete, with subscripts to keep track of which one is which: $X_{AB}, X_{Ab}, X_{aB},$ and X_{ab}.

What is the probability of a random gamete being *AB*? Assuming that there is a random association of the two loci, this is simply the probability of having the *A* allele *and* having the *B* allele, which is obtained by multiplying the frequency of the A allele by the frequency of the B allele, which is $p_1 p_2$. Likewise, the expected frequency of gamete *Ab* would be the probability of having the *A* allele and the *b* allele, which is $p_1 q_2$. Figure 2.2 shows all the expected frequencies for the two loci, and also the expected haplotype frequencies under the assumption that there is a random association between the alleles.

Figure 2.2 shows the equilibrium situation. Just as with Hardy–Weinberg equilibrium for a single locus, nonrandom mating and the evolutionary forces can lead to a deviation from the random expectation. When this occurs, the loci are considered at a state of **linkage disequilibrium** (also known as *gametic disequilibrium*). Some haplotypes occur more frequently than expected, and some occur less frequently than expected. Linkage disequilibrium is measured by a quantity known as *D*; the effects of linkage disequilibrium are shown in Figure 2.3, where some haplotypes are increased by the quantity *D* and some are reduced. Note that the equilibrium case is defined as $D = 0$. *D* can be estimated from any of the boxes shown in Figure 2.3. For example, $D = X_{AB} - p_1 p_2$. Similar computation can be made from any of the boxes in Figure 2.3 and they will all give the same value of *D*. In actual research, there are various ways to test the null hypothesis of $D = 0$.

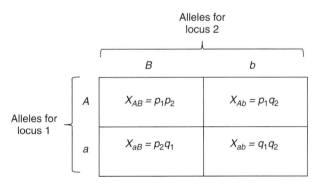

FIGURE 2.2 Expected distribution of haplotype frequencies for two loci, each with two alleles. *X* represents the haplotype frequencies under the assumption of a random association of alleles. For example, X_{Ab} results when one gamete contains the *A* allele for the first locus and the *b* allele from the second locus. For locus 1, p_1 is the frequency of the *A* allele and q_1 is the frequency of the *a* allele. For locus 2, p_2 is the frequency of the *B* allele and q_2 is the frequency of the *b* allele.

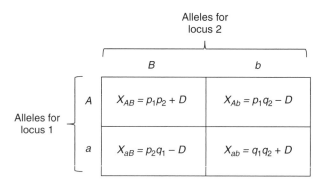

FIGURE 2.3 Expected distribution of haplotype frequencies for two loci, each with two alleles. X represents the haplotype frequencies allowing for linkage disequilibrium, the nonrandom association of alleles. D represents the amount of linkage disequilibrium. All other symbols are as defined in Figure 2.1. Note that under equilibrium, $D = 0$ and the table reduces to the form shown in Figure 2.2.

We saw earlier that if a population is not initially at Hardy–Weinberg equilibrium for a single locus, it will reach equilibrium within a single generation. This is not the case for linkage disequilibrium. If the evolutionary force(s) that led to linkage disequilibrium is (are) no longer in effect and there is random mating, then recombination will eventually shuffle the haplotype frequencies and an equilibrium state ($D = 0$) will be reached. However, unlike the single-locus case, it will take some time for this to happen. The amount of time is dependent on the rate of recombination and the number of generations that have passed since the initial state of linkage disequilibrium. The decay of linkage disequilibrium over time is useful in a number of contexts to allow detection of evolutionary change.

B. More than Two Alleles

Although we have been using a simple model of a single locus with two alleles, Hardy–Weinberg equilibrium can easily be extended to more than two alleles. For example, imagine that we have three alleles (A, B, and C) with p as the frequency of the A allele, q as the frequency of the B allele, and r as the frequency of the C allele. Applying the rules of probability used earlier, we can figure out the expected genotype frequencies under Hardy–Weinberg equilibrium as follows:

$$AA = p^2$$
$$AB = 2pq$$
$$AC = 2pr$$
$$BB = q^2$$
$$BC = 2qr$$
$$CC = r^2$$

The same method can be used on any number of alleles at a single locus.

C. X-Linked Genes

The examples used in this chapter all consider a bisexual species (two sexes). To make the models easier to use, we have implicitly assumed that the allele frequencies are the same in males and females. This assumption does not hold true for all loci on the X chromosome of the sex chromosomes. Remember that females have two X chromosomes whereas males have only one X chromosome (the other one is the Y chromosome). Because the Y chromosome is smaller than the X chromosome, this means that some sequences on the X chromosome have no corresponding (homologous) sequence on the Y chromosome. As such, any recessive genes on the X chromosome will be expressed in males with only one copy of the X chromosome, whereas females will need two X chromosomes to express the trait.

The Xg blood group system is used here to demonstrate the different patterns of sex-linked inheritance. This locus is found on the X chromosome and has two alleles, the dominant allele Xg^a, which codes for the Xg(a+) blood type, and the recessive allele Xg, which codes for the Xg(a−) blood type (Sanger et al. 1971). In females, who inherit two copies of the X chromosome, there are three genotypes and two phenotypes:

Genotype	Phenotype
$Xg^a Xg^a$	Xg(a+)
$Xg^a Xg$	Xg(a+)
$Xg Xg$	Xg(a−)

This list is the typical correspondence of genotypes and phenotypes with one dominant allele and one recessive allele. In males, however, the situation is different because the male inherits only one copy, as there is no corresponding gene on the Y chromosome. If the male inherits the Xg^a allele from his mother, he will have Xg(a+) type blood, and if he inherits the Xg allele from his mother, he will have Xg(a−) type blood. There are two genotypes and two phenotypes:

Genotype	Phenotype
Xg^a	Xg(a+)
Xg	Xg(a−)

There are different Hardy–Weinberg equilibrium proportions for males and females for an X-linked genetic marker. Given two alleles, A and a, with frequencies p and q, respectively, the genotype frequencies for females are those expected under Hardy–Weinberg equilibrium following the usual distribution: $AA = p^2, Aa = 2pq, aa = q^2$. For males, there are two genotypes, A and a, corresponding to the allele that is on the X chromosome from the mother, as the Y chromosome from the father does not contribute to the genotype. The Hardy–Weinberg equilibrium frequencies are equal to the allele frequencies; the frequency of genotype A is p and the frequency of genotype a is q. As an example, consider the allele frequencies for the Xg blood group among northern Europeans: $Xg^a = 0.659$ and $Xg = 0.341$

TABLE 2.2 Hardy–Weinberg Equilibrium Genotype and Phenotype Frequencies for a Sex-Linked Genetic Marker, the Xg Blood Group System[a]

Sex	Genotype	Frequency	Phenotype	Frequency
Females	$Xg^a Xg^a$	$p^2 = 0.4343$	Xg(a+)	0.8837
	$Xg^a Xg$	$2pq = 0.4494$	Xg(a+)	
	$XgXg$	$q^2 = 0.1163$	Xg(a−)	0.1163
Males	Xg^a	$p = 0.6590$	Xg(a+)	0.6590
	Xg	$q = 0.3410$	Xg(a−)	0.3410

[a]The data presented in this table are based on allele frequencies from northern Europe (Sanger et al. 1971) of $Xg^a = 0.659$ and $Xg = 0.341$.

(Sanger et al. 1971). The genotype frequencies under Hardy–Weinberg equilibrium are presented in Table 2.2 for males and females. Note that the proportion of individuals with Xg(a−) type blood is almost 3 times higher in males (0.341) than in females (0.116). This sex difference is attributed to the fact that males need only one copy of the Xg allele, whereas females need two, which is a much lower probability: q^2 rather than q.

VI. HARDY–WEINBERG EQUILIBRIUM AND EVOLUTION

Although this chapter has probably seemed rather abstract in many places, and certainly more mathematical than anthropological, a consideration of Hardy–Weinberg equilibrium is needed to get into the specifics of microevolution. The rest of this book builds on the baseline model of Hardy–Weinberg equilibrium.

A population might not be at Hardy–Weinberg equilibrium because of deviations due to nonrandom mating (inbreeding, assortative mating) and/or action of one or more of the evolutionary forces (mutation, natural selection, genetic drift, gene flow). An importance difference is that nonrandom mating causes changes in genotype frequencies but not allele frequencies, whereas the evolutionary forces lead to changes in both genotype and allele frequencies. We start with a consideration of the genotypic impact of inbreeding as a form of nonrandom mating in the next chapter, and then move on to separate chapters on the different evolutionary forces. These different factors will be discussed initially one at a time to get the concept across, and then combined where appropriate.

Keep in mind that although this approach focuses on one factor at a time, in reality *all* of these deviations from Hardy–Weinberg equilibrium can be operating at the same time. As I often tell my students in my introductory class, the good news is that there are only four evolutionary forces, making exam review easier. However, the bad news as far as researchers are concerned is that these forces (along with nonrandom mating) can interact in many, often very complex, ways. It becomes a problem (or challenge, depending on how you look at it) to untangle these effects, particularly when dealing with the culturally complex, highly mobile, and rapidly changing human species.

VII. SUMMARY

Genes and DNA sequences have different forms, known as *alleles*. When we look at microevolution, we are interested in how the relative frequency of alleles can change over time. When the field of genetics was just beginning, some researchers wondered why dominant alleles did not increase to completely replace recessive alleles over time. This question led to a deeper understanding of the mathematical application of Mendel's ideas on a populational basis. In 1908, mathematician Godfrey Hardy and physician Wilhelm Weinberg independently derived a mathematical relationship between allele and genotype frequencies, and further showed that, under certain conditions, the allele and genotype frequencies would remain the same generation after generation. For a locus with two alleles, A and a, with allele frequencies p and q, respectively, the genotype frequencies in the next generation will be $AA = p^2, Aa = 2pq$, and $aa = q^2$. Under certain conditions, the allele and genotype frequencies will remain the same over time.

This principle, known as **Hardy–Weinberg equilibrium**, provides a baseline for understanding evolutionary change. Because the equilibrium model makes a number of assumptions, any actual change in genotype and allele frequencies means that one or more of these assumptions are not met. By comparing conditions in the real world to the theoretical assumptions of Hardy–Weinberg equilibrium, we can develop insight into the evolutionary process. One assumption of Hardy–Weinberg equilibrium is random mating. When this assumption is not met, as will occur under inbreeding, the genotype frequencies are changed although there is no change in allele frequency.

Consideration of the assumptions of Hardy–Weinberg equilibrium shows that there are four forces of evolution, which lead to allele frequency change (microevolution) over time: mutation, natural selection, genetic drift, and gene flow. *Mutation* is a random change in the genetic code, resulting in the introduction of new alleles into a population at low frequency. The other three evolutionary forces act on mutations, causing the frequency to increase or decrease. *Natural selection* occurs when individuals with different genotypes have different probabilities of survival and/or reproduction. Individuals who are more likely to survive and reproduce pass on their alleles to the next generation in greater numbers, altering allele frequencies over time. *Genetic drift* is sampling variation, such that allele frequencies will fluctuate randomly over time, sometimes increasing and sometimes decreasing. The degree of expected genetic drift is dependent on population size; smaller populations show more drift. *Gene flow* is the movement of alleles from one population to another accompanying migration of individuals. Gene flow can introduce new alleles into a population, and acts to reduce genetic differences between populations.

APPENDIX 2.1 PROOF SHOWING HOW ALLELE FREQUENCIES CAN BE COMPUTED FROM GENOTYPE FREQUENCIES

Assume a locus with two alleles, A and a, and three genotypes: AA, Aa, and aa. Let N_{AA}, N_{Aa}, and N_{aa} represent the numbers of individuals with genotypes AA, Aa, and *aa*, respectively. Let N be the total number of all genotypes:

$$N = N_{AA} + N_{Aa} + N_{aa}$$

The genotype frequencies are obtained by dividing the number of each genotype by the total number of all genotypes, giving

$$f_{AA} = \frac{N_{AA}}{N}$$

$$f_{Aa} = \frac{N_{Aa}}{N}$$

$$f_{aa} = \frac{N_{aa}}{N} \qquad \text{(A2.1)}$$

Using the allele counting method, we find that the number of A alleles is equal to

$$2N_{AA} + N_{Aa}$$

because we count the homozygote (AA) twice. We then divide this number by the total number of alleles in the population, which is twice the number of people because each person has two alleles. This gives

$$p = \frac{2N_{AA} + N_{Aa}}{2N}$$

We can express this quantity as the sum of two fractions as

$$p = \frac{2N_{AA}}{2N} + \frac{N_{Aa}}{2N}$$

which reduces to

$$p = \frac{N_{AA}}{N} + \frac{N_{Aa}}{2N}$$

Using the formulas for genotype frequencies f_{AA} and f_{Aa} from equation (A2.1) gives

$$p = f_{AA} + \frac{f_{Aa}}{2}$$

In similar manner, the frequency of the *a* allele is

$$q = \frac{N_{Aa} + 2N_{aa}}{2N}$$

which gives

$$q = \frac{f_{Aa}}{2} + f_{aa}$$

The derivation for p and q above lead to equation (2.3) in the main (chapter) text.

APPENDIX 2.2 USING THE CHI-SQUARE STATISTIC TO TEST FOR HARDY–WEINBERG EQUILIBRIUM

As shown earlier in this chapter, we can compare observed genotype numbers with the numbers expected under Hardy–Weinberg equilibrium. This allows us to test the hypothesis that the population is at Hardy–Weinberg equilibrium (for the specific locus). We use a chi-square test, a method that is used for comparing observed and expected numbers in statistical research. The chi-square statistic, denoted by the symbol χ^2, is computed using the formula

$$\chi^2 = \sum \frac{(O - E)^2}{E} \tag{A2.2}$$

Here, O refers to the observed number for a given genotype and E refers to the expected number for a given genotype. The difference between observed and expected $(O - E)$ is squared and divided by the expected number. The Greek letter sigma ($\sum$) is used in statistics as shorthand for summation; in this case, it means performing the same operations on each genotype and then summing the results.

The computation of the chi-square statistic is presented below for one of the examples presented in the text. The first three columns of the table show the genotype, the observed number, and the expected number. The remaining columns walk us through the computation of the chi-square statistic. As shown in the main text, the allele frequencies $p = 0.7$ and $q = 0.3$ were computed from the observed genotype numbers and used to compute the expected number (expected under Hardy–Weinberg equilibrium):

Genotype	Observed Number (O)	Expected Number (E)	$(O - E)$	$(O - E)^2$	$(O - E)^2/E$
AA	40	39.2	0.8	0.64	0.0163
Aa	32	33.6	−1.6	2.56	0.0762
aa	8	7.2	0.8	0.64	0.0889

The chi-square statistic is now computed using equation (A2.2), where the values in the last column of the table are summed:

$$\chi^2 = 0.0163 + 0.0762 + 0.0889 = 0.1814$$

This value needs to be compared with a standard table of chi-square values for a given probability value, which is the level of statistical significance that we use for testing the hypothesis. A full discussion of this concept is beyond the scope of this text, and you might want to consult a statistics text. We often tend to use a level of 5% (0.05) to test hypotheses. The final thing we need to know is a statistic known as the *degrees of freedom*, which represents the number of values that are free to vary. For a chi-square test, the degree(s) of freedom, symbolized as df, is computed as

$$df = n - k - 1 \tag{A2.3}$$

where n is the number of classes and k is the number of independent parameters. We have three genotypes, so $n = 3$. The genotype frequencies are computed from the allele frequencies, p and q, which are the parameters. However, if we know either p or q, we can always compute the other one (because $p + q = 1$), so there is only one independent parameter, and therefore $k = 1$. This means that for the examples discussed in this chapter, equation (A2.3) gives df $= 3 - 1 - 1 = 1$. A table of chi-square values, available in almost any statistics book or online, shows that for a significance level of 0.05 and df $= 1$, the critical value of the chi-square statistic is 3.841.

Now we can perform the hypothesis test. If the observed chi-square statistic is less than the critical value, then we accept the null hypothesis that the population is at Hardy–Weinberg equilibrium. On the other hand, if the observed chi-square statistic is greater than or equal to the critical value, then we reject the hypothesis of Hardy–Weinberg equilibrium. In the example above, the observed chi-square value is 0.1814, which is *less* than the critical value of 3.841. Therefore, the slight difference that we see between observed and expected genotype numbers is *not* statistically significant and the population is at Hardy–Weinberg equilibrium.

Let us consider the second example used in this chapter. As shown in the main text, the allele frequencies $p = 0.8$ and $q = 0.2$ were computed from the observed genotype numbers and used to compute the expected number (expected under Hardy–Weinberg equilibrium):

Genotype	Observed Number (O)	Expected Number (E)	$(O - E)$	$(O - E)^2$	$(O - E)^2/E$
AA	90	83.2	6.8	46.24	0.5558
Aa	28	41.6	−13.6	184.96	4.4462
aa	12	5.2	6.8	46.24	8.8923

After performing the calculations in this table, we get a chi-square (χ^2) statistic of

$$\chi^2 = 0.5558 + 4.4462 + 8.8923 = 13.8943$$

This value is *much* larger than the critical value of 3.841, so we reject the hypothesis of Hardy–Weinberg equilibrium.

The chi-square test is most appropriate for large samples. For small samples, other methods are often used to test for Hardy–Weinberg equilibrium (Mielke et al. 2011).

CHAPTER 3

INBREEDING

Paradoxes can be fun and/or frustrating, but are also useful in illustrating the illogical outcomes that sometimes follow invalid assumptions. Consider the following paradox that arises from considering the process of genetic inheritance extended into the past. As noted in Chapter 1, you have two ancestors one generation in the past—your biological parents. Because each parent also has two parents, this means that you have four grandparents. Extending this back means that you had eight great-grandparents, 16 great-great grandparents, and so on. In mathematical terms, we can express the number of ancestors as 2^n (2 raised to the nth power), where n is the number of generations in the past. The number of ancestors thus increases exponentially into the past. After 5 generations, you have $2^5 = 32$ ancestors, and after 10 generations you have $2^{10} = 1024$ ancestors. The number of ancestors increases rapidly thereafter; at 30 generations you have over one billion ancestors (1,073,741,824), and at 40 generations you have over one trillion ancestors ($2^{40} = 1,099,511,627,776$)! This is not that long ago in the past; given an average of 25 years per generation, 40 generations is only 1000 years ago.

This last number is preposterously large and illustrates the paradox—there were not that many people living in the world 1000 years ago. Indeed, there were only 6.9 billion people in the world as of mid-2010 (Population Reference Bureau 2010). Clearly, something is wrong with our simple model of each person having two parents. The paradox is resolved by noting that the simple exponential model (2^n) assumes that *each* ancestor is unique. In reality, some ancestors are not unique, but are shared by more than one line of descent, which is indicative of inbreeding.

Comparison of Figures 3.1 and 3.2 provides a simple illustration of inbreeding. Figure 3.1 shows the genealogy of a hypothetical individual (Amy) going back three generations (a genealogy is also known as a *pedigree*). In this example, Amy has two parents, four grandparents, and eight great grandparents as expected from the exponential model 2^n. Each ancestor is unique. Figure 3.2 presents a different genealogy of a hypothetical individual (Alice) for three generations, where there are four unique grandparents but only seven unique great grandparents. In this second case, one woman (Doris) has children with two different men (Davis and Desmond). This means that Alice's parents in Figure 3.2 (Boris and Barbara) are related (they are half first cousins), and that Alice is inbred. Specifically,

Human Population Genetics, First Edition. John H. Relethford.
© 2012 Wiley-Blackwell. Published 2012 by John Wiley & Sons, Inc.

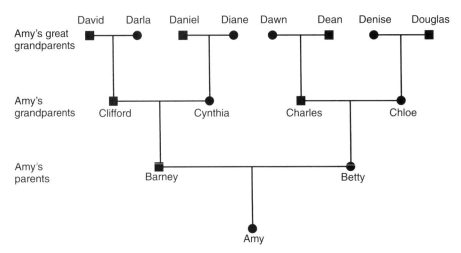

FIGURE 3.1 Genealogy of a reference person (Amy) going back three generations. Squares represent males; circles, females. Horizontal lines indicate matings, and vertical lines indicate descent from these matings. Note that the number of Amy's ancestors doubles with each generation into the past.

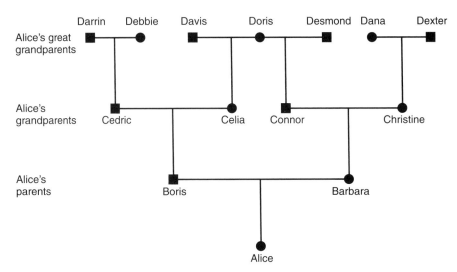

FIGURE 3.2 Genealogy of an inbred reference person (Alice) going back three generations. Squares indicate males; circles, females. Horizontal lines indicate matings, and vertical lines indicate descent from these matings. Note that in this example, Alice has only seven great grandparents (compared with eight in Figure 3.1). Alice's father's mother's mother is the same as her mother's father's mother (Doris). This makes Alice's parents (Boris and Barbara) half first cousins, because they share one grandparent (Doris).

Alice's father's mother's mother and her mother's father's mother are the same person (Doris). Ultimately, everyone is inbred to some extent, because of the mathematical impossibility of having two unique ancestors for every person as you trace a genealogy into the past. Instead, lines of descent combine multiple times in the past.

As noted in the previous chapter, Hardy–Weinberg equilibrium assumes that all mating is random and that two mates are completely unrelated; in other words, no inbreeding. We can see that this will never be strictly true, as we are all related and inbred to some extent. However, the genetic impact of inbreeding depends on the closeness of the relationship of the parents. If two parents are seventh cousins, then the degree of relationship is so small that for all practical purposes they are unrelated. More significant deviations from the assumption of random mating are associated with close degrees of relationship, such as mating between siblings, uncles and nieces (or aunts and nephews), or first cousins.

In addition, as noted earlier, inbreeding does not directly change the frequency of an allele, and is therefore not an evolutionary force. Instead, inbreeding changes the genotype proportions from those expected under Hardy–Weinberg equilibrium, such that the frequency of homozygotes is increased, and the frequency of heterozygotes is decreased. Although inbreeding by itself does not change *allele* frequencies, it can interact with one of the evolutionary forces to change the *rate* of allele frequency change. As an example (to be covered in more detail in Chapter 6), imagine a case where there is selection against a recessive homozygote. Under inbreeding, there will be a greater number of these homozygotes with each generation such that a greater proportion of the population will be selected against, and the speed of selection will therefore be greater than under Hardy–Weinberg equilibrium. Again, the exact impact of inbreeding relative to random mating will depend on the level of inbreeding.

This chapter begins with the concept of the inbreeding coefficient (a measure of the level of inbreeding) and how inbreeding is determined from genealogical data. This introduction is followed by an explanation of how inbreeding affects genotype frequencies but not allele frequencies. This chapter concludes with discussion of variation in rates of inbreeding in human populations as well as several case studies of measurement and analysis of inbreeding in human populations. Inbreeding is of particular interest in studies of human population genetics because different cultures may have different rules and preferences for and against different levels of inbreeding. The study of inbreeding in human populations thus provides an opportunity to examine how human culture affects genetic variation.

I. QUANTIFYING INBREEDING

As noted earlier, inbreeding is the result of mating between relatives. However, as was just discussed, it is clear that we are all inbred to some extent or another because everyone's ancestry becomes commingled with everyone else's. Genetically, most of our concern is with relatively *close* levels of inbreeding, which translates practically into a concern with common ancestry within a handful of generations. Questions about inbreeding are questions about common ancestry, and are best explained by examining the genealogical relationship between individuals.

A. Genealogies and Inbreeding

Genealogical diagrams provide the best way to understand common ancestry and inbreeding. There are several different ways of drawing genealogies, with different symbols used to designate males and females, and different arrangements of lines to illustrate lines of descent. How would you draw your family tree back three generations? One way of doing this has already been shown in Figures 3.1 and 3.2, with horizontal lines indicating matings and vertical lines to indicate descent. Another method is shown in Figure 3.3, depicting the same relationships as in Figure 3.2 but eliminating the horizontal lines and using diagonal lines to indicate descent. This method is used throughout this text because it is easier to see descent and genetic relationships when discussing inbreeding.

Because our focus here is on *genetic* relationships, all of the hypothetical genealogies used here refer to *mating* (a biological act) and not *marriage* (a culturally defined relationship). Although anthropologists and geneticists will sometimes use marriage records to reconstruct genealogies, we have to keep in mind the possibility that some births result from extramarital relationships. Likewise, a genealogy based on cultural definitions of family and kinship could include relationships that were established through adoption or remarriage (stepparents), which would not indicate an actual genetic relationship. Unless otherwise noted, all discussions of inbreeding here refer to one's *biological* relatives.

Figures 3.2 and 3.3 both show an inbred person whose parents are half first cousins. The nature of inbreeding is very clear from Figure 3.3. In any genealogy, a person is inbred if you can draw a line back through to a common ancestor and

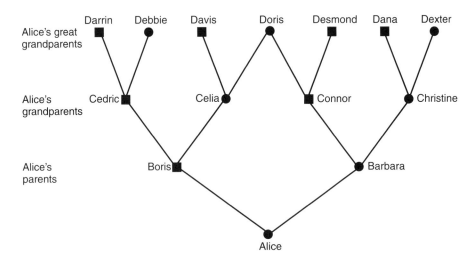

FIGURE 3.3 Genealogy of an inbred reference person (Alice) going back three generations. Squares represent males; circles, females. This diagram shows the same inbreeding relationship as in Figure 3.2, but diagonal lines are used to indicate descent from pairs of parents. Although the system used in Figures 3.1. and 3.2 is often preferred in genealogical illustration, the system of diagonal lines is used throughout the text, as it is easier to convey descent and genetic inheritance.

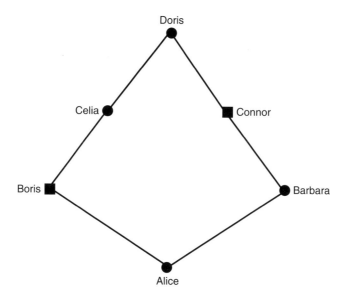

FIGURE 3.4 Genealogy of an inbred person (Alice) showing only the lines of descent that contribute to inbreeding (all other individuals are not shown). This figure is the same as Figure 3.3 after eliminating all ancestors who are not involved in the inbreeding. As shown here, the nature of inbreeding as a "loop" of ancestry is clear. The complete loop is Alice–Boris–Celia–Doris–Connor–Barbara–Alice.

then back to the original person through a different line. In other words, a person is inbred if you can draw a loop back through one or more common ancestors. This loop is even clearer in Figure 3.4, which takes Figure 3.3 and eliminates those ancestors who are not part of the inbreeding loop. Figure 3.4 shows the loop as starting with Alice and then going backward three generations to the common ancestor (Doris). The complete loop is Alice–Boris–Celia–Doris–Connor–Barbara–Alice.

B. Types of Inbreeding

We can use this simplified form of genealogy, showing only ancestors in the chain of inbreeding, to illustrate different types of inbreeding. Extremely close forms of inbreeding occur with the offspring of parent–child mating, sib mating, and half-sib mating, all of which are shown in Figure 3.5. Different types of cousin marriage are shown in Figure 3.6. Cousins are referred to by number—first, second, third, and so on—which refers to the depth of the relationship. First cousins share one or more grandparents; the diagram in Figure 3.6 shows *full first cousins*, who share two grandparents. Second cousins share great grandparents, and third cousins share great-great grandparents.

There are also variations on the different types of cousin marriages. As shown in Figure 3.7, "first cousins once removed" describes the relationship between a person and the child of their first cousin. The "removed" term refers to the number of generations apart the two cousins are separated; "once removed" is one generation apart, "twice removed" refers to two generations apart, and so on.

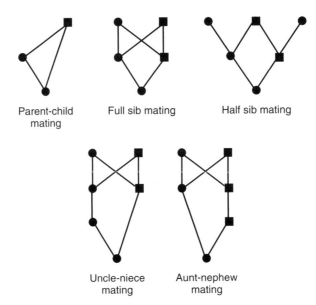

Parent-child   Full sib mating   Half sib mating
mating

Uncle-niece   Aunt-nephew
mating   mating

FIGURE 3.5 Diagrams of five types of close inbreeding. As in previous figures, only ancestors that are part of the loop through common ancestors are shown.

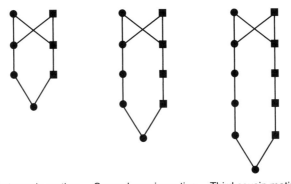

First cousin mating   Second cousin mating   Third cousin mating

FIGURE 3.6 Diagrams of cousin mating. As in previous figures, only ancestors that are part of the loop through common ancestors are shown. First cousins share a set of grandparents, second cousins share a set of great grandparents, and third cousins share a set of great grandparents.

It might be useful to think about your own family to understand the relationship between various cousins. If you have a first cousin, think of that person and the set of grandparents that you have in common with that cousin. If your cousin has a child, that child is your first cousin once removed. If you also have a child, your child and your first cousin's child are second cousins. Likewise, if your second cousin has a child, that child is your second cousin once removed. If that child then has a child, your second cousin's grandchild will be your second cousin twice removed. Yes, it can get very complicated quickly!

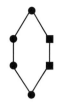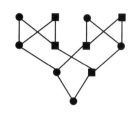

First cousins once   Half first cousin mating Double first cousin mating
removed mating

FIGURE 3.7 Diagrams of variations of first cousin mating. As in previous figures, only ancestors that are part of the loop through common ancestors are shown. Mating between first cousins once removed is between a person and the child of their first cousin, thus involving a mating between generations. First cousins share two grandparents, as shown in Figure 3.6. In half first cousin mating, only one grandparent is shared. In double first cousin mating, *all* four grandparents are shared. Note that half first cousin mating was already shown, albeit in somewhat different form, in Figures 3.3 and 3.4.

Figure 3.7 shows two other variations on first cousin mating. Whereas first cousins as shown in Figure 3.6 share two grandparents, half first cousins share only one grandparent (as was shown earlier in Figures 3.2 and 3.3). The last example in Figure 3.7 shows the mating between double first cousins. Two people are double first cousins if they share *all* of their grandparents. As shown here, picture two couples, each having a male child and a female child, and then having the brother and sister in one family mating with the brother and sister in the other family, producing double first cousins.

All of the matings shown in Figures 3.5–3.7 produce a child that is inbred, because all of these diagrams show a loop back through one or more common ancestors. Note that while it only takes one common ancestor to define a case of inbreeding, some types of mating involve more than one common ancestor, and thus we can draw more than one loop. For example, in Figure 3.6, full first cousin mating involves two common ancestors, and we could draw two loops, one for each grandparent. Half first cousin mating involves only one common ancestor, and we can draw only one loop. Double first cousin mating has four common ancestors (all of the grandparents), and we can therefore draw four loops (see Figure 3.7).

C.   The Inbreeding Coefficient

We will now use the concept of inbreeding as a genealogical loop through one or more common ancestors to provide a quantitative measure of inbreeding known as the **inbreeding coefficient**. As noted in Chapter 1, *homozygosity* refers to the case where an individual has inherited the same allele from both parents. In the simple example of a locus with two alleles, *A* and *a*, the homozygous genotypes are *AA* and *aa*. Homozygosity is defined as the two alleles being identical, but there are actually two reasons why this identity can occur—identity by descent and identity by state. **Identity by descent** is when the two alleles are the same because they were both inherited from a common ancestor. **Identity by state** occurs when the identical alleles do *not* come from a common ancestor. The inbreeding coefficient (typically denoted by the symbol *F*) is the probability that an inbred individual has two alleles at a given locus that are identical because of inheritance from a common ancestor. *F* is thus the probability of identity by descent.

How to Compute the Inbreeding Coefficient

Figure 3.8 shows how this probability can be computed from the genealogy of an inbred individual. In this case, we are using the case of a person whose parents

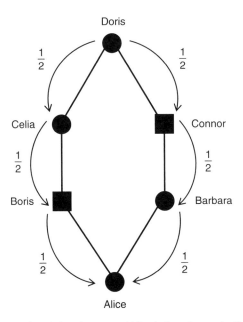

FIGURE 3.8 Genealogy of an inbred person (Alice) showing only the lines of descent that contribute to inbreeding (all other individuals are not shown). This is the same genealogy as in Figure 3.4. The curved arrows indicate the path of genetic descent with the associated probability of $\frac{1}{2}$ from parent to child. These probabilities allow us to compute the probability of Alice having a homozygous genotype (*AA* or *aa*) because of identity by descent. Imagine that the common ancestor (Doris) has the genotype *Aa*. The probability of her passing on the *A* allele to her daughter Celia is $\frac{1}{2}$. The probability of Celia passing that *A* allele on to Boris is also $\frac{1}{2}$, as is the probability of Boris passing it on to Alice. Thus, the probability of the *A* allele being passed on from Doris to Celia to Boris to Alice is computed using the AND rule as $\frac{1}{2} \times \frac{1}{2} \times \frac{1}{2} = \frac{1}{8}$. Now, for Alice to receive an *A* allele from the other line of descent from the common ancestor Doris, the probability of the allele being passed from Doris to Connor AND from Connor to Barbara AND from Barbara to Alice is also $\frac{1}{2} \times \frac{1}{2} \times \frac{1}{2} = \frac{1}{8}$. In order for Alice to receive an *A* allele from *both* lines of descent from Doris is the probability of both events occurring, which is $\frac{1}{8} \times \frac{1}{8} = \frac{1}{64}$. We also have to consider the probability of Alice having the other homozygous genotype (*aa*) because of identity by descent, which would occur when Alice receives the *a* allele from both lines of descent from Doris. This probability would also be $\frac{1}{64}$. Because the probability of Alice having the *AA* genotype because of identity by descent is $\frac{1}{64}$, and the probability that Alice will have the *aa* genotype because of identity by descent is also $\frac{1}{64}$, the probability of her having either the *AA* or the *aa* homozygous genotype is solved using the OR rule as $\frac{1}{64} + \frac{1}{64} = \frac{2}{64} = \frac{1}{32}$. Thus, $F = \frac{1}{32} = 0.03125$.

were half first cousins as was shown previously in Figures 3.3 and 3.4. The curved lines indicate the transmission of alleles from parent to child across the generations; in each case, the probability of a parent passing on a given allele to a child is $\frac{1}{2}$. As described in the legend to Figure 3.8, the probability of Alice getting the A allele from both lines of descent from the common ancestor Doris is $\frac{1}{64}$, and the probability of obtaining the a allele from both lines of descent is also $\frac{1}{64}$. For Alice to have *either* homozygous genotype (AA or aa) is computed using the OR rule, giving $F = \frac{1}{64} + \frac{1}{64} = \frac{2}{64} = \frac{1}{32} = 0.03125$.

An easier shortcut can be used here:

$$F = \left(\tfrac{1}{2}\right)^i$$

where i is the number of individuals that lie in the loop up through the common ancestor and then down again, not counting the inbred person. Referring to Figure 3.8, we can identify the ancestors that lie on the loop from Alice to Doris and back (not counting Alice) as Boris–Celia–Doris–Connor–Barbara. There are five people in this loop, so $i = 5$ and $F = \left(\frac{1}{2}\right)^5 = \frac{1}{32} = 0.03125$. Thus, the probability of Alice having a homozygous genotype (AA or aa) because of identity by descent is 0.03125.

Computation of F gets a little bit more complicated when there is more than one common ancestor, because we need to add up the probabilities over all possible paths of descent. Consider the case of first cousin mating shown in Figure 3.9. There are two common ancestors, numbered 6 and 7. We therefore need to consider two inbreeding loops. One loop is through ancestor number 6: 1–2–4–6–5–3–1. The other loop goes through common ancestor number 7: 1–2–4–7–5–3. We now compute the inbreeding coefficient for each loop, and then add them together. The method of dealing with more than one common ancestor can be summarized with the following formula:

$$F = \sum \left(\tfrac{1}{2}\right)^i \tag{3.1}$$

The symbol $\sum$ is the uppercase Greek letter sigma, which is used in mathematics to *sum* whatever terms are to the right of the symbol. In this case, the summation is over the number of common ancestors. In Figure 3.9, there are two common ancestors—individuals 6 and 7. For ancestor 6, we have $\left(\frac{1}{2}\right)^i = \left(\frac{1}{2}\right)^5$, because there are five ancestors in the loop through this ancestor. For ancestor 7, the value $\left(\frac{1}{2}\right)^i$ is also $\left(\frac{1}{2}\right)^5$ because there are five ancestors in the loop through ancestor 7. Using equation (3.1), the total inbreeding coefficient is therefore the sum of these, giving $\frac{1}{32} + \frac{1}{32} = \frac{1}{16} = 0.0625$.

Inbreeding Coefficients for Different Levels of Relationship
Table 3.1 presents inbreeding coefficients for various types of relationship ranging from parent–child through third cousin matings; most of the matings in Table 3.1 are shown in Figures 3.5–3.7. To make sure that you understand how to compute the inbreeding coefficient, you should try using equation (3.1) on the different genealogies in Figures 3.5–3.7 to see if you get the same values.

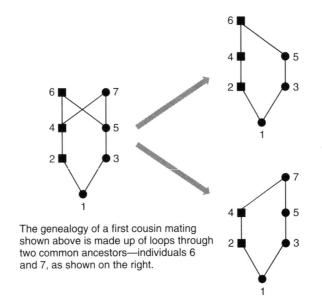

The genealogy of a first cousin mating shown above is made up of loops through two common ancestors—individuals 6 and 7, as shown on the right.

FIGURE 3.9 Computing the inbreeding coefficient of first cousin mating. The genealogy on the left side represents the mating of first cousins (individuals 2 and 3). There are two common ancestors, labeled here as individuals 6 and 7. In order to compute the inbreeding coefficient for individual 1, we need to count the number of ancestors in each loop shown on the right side of the figure. For common ancestor 6, the number of ancestors in the loop (not counting individual 1) is $i = 5$. For common ancestor 7, the number of ancestors is also $i = 5$. Using equation (3.1), the inbreeding coefficient for a first cousin mating is $F = \sum (\frac{1}{2})^i = (\frac{1}{2})^5 + (\frac{1}{5})^5 = \frac{1}{32} + \frac{1}{32} = \frac{2}{32} = \frac{1}{16} = 0.0625$.

Keep in mind that inbreeding can also be cumulative when a person's parents are related in more than one way. Figure 3.10 presents a genealogy where person 1 is inbred because her parents (individuals 2 and 3) are simultaneously first cousins and second cousins. Listing all possible loops of common ancestry and using equation (3.1), we compute the inbreeding coefficient of person 1 as $F = 0.078125$.

Dealing with an Inbred Common Ancestor
All of the examples given thus far have assumed that the common ancestors are themselves not inbred. In reality, one or more common ancestors may be inbred, which will act to increase the inbreeding coefficient of descendants. Figure 3.11 shows an example of a person (individual 1) whose parents are first cousins. However, unlike previous examples of first cousin mating, this example shows that one of the common ancestors (person 6) was also a child of a first cousin mating. The computation of the inbreeding coefficient can be extended to handle cases like this by using the formula

$$F = \sum \left(\tfrac{1}{2}\right)^i (1 + F_A) \qquad (3.2)$$

TABLE 3.1 Inbreeding Coefficients for Different Types of Mating

Mating	Inbreeding Coefficient) (F)
Parent–child	$\frac{1}{4} = 0.25$
Full sibs	$\frac{1}{4} = 0.25$
Half sibs	$\frac{1}{8} = 0.125$
Uncle–niece/aunt–nephew	$\frac{1}{8} = 0.125$
Double first cousins	$\frac{1}{8} = 0.125$
First cousins	$\frac{1}{16} = 0.0625$
Half first cousins	$\frac{1}{32} = 0.03125$
First cousins once removed	$\frac{1}{32} = 0.03125$
Second cousins	$\frac{1}{64} = 0.015625$
Second cousins once removed	$\frac{1}{128} = 0.0078125$
Third cousins	$\frac{1}{256} = 0.00390625$

where F_A is the inbreeding coefficient of the common ancestor in any given path, and these results are summed over all possible paths. An example of computation is presented in the legend for Figure 3.11. When common ancestors are themselves inbred, the inbreeding coefficient increases. When computing the inbreeding coefficient from genealogical data, we often quickly run into a point back in time beyond which we have no further information, and necessarily have to assume that the common ancestors were not inbred. This means that we are likely to underestimate the true value of F because of incomplete genealogies.

The cumulative effect of continued inbreeding over many generations can sometimes be quite substantial. An example of this comes from Alvarez et al.'s (2009) study of inbreeding in the Spanish Hapsburg (also known as Habsburg) dynasty from 1516 to 1700. In order to maintain control of their line, many marriages between close relatives occurred in this dynasty, including uncle–niece and cousin marriages. In addition, their analysis showed a considerable amount of accumulated inbreeding due to common ancestors being inbred. The inbreeding coefficient of six Hapsburg kings ranged from 0.025 to 0.254 with an average value of 0.129, an extremely high value for human populations! Close analysis of the pedigrees shows that inbreeding levels in these kings increased greatly by the impact of multiple remote ancestors.

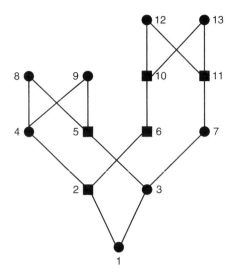

FIGURE 3.10 Example of a genealogy where the parents of an inbred person (individual 1) are both first and second cousins. As in previous figures, only ancestors directly in one or more loops of inbreeding are shown. The parents (individuals 2 and 3) are simultaneously first cousins (through ancestors 8 and 9) and second cousins (through ancestors 12 and 13). In order to compute the inbreeding coefficient, we need to tally the number of ancestors (*i*) in each possible loop through a common ancestor. The loops and number of ancestors in each loop are as follows:

$$1 - 2 - 4 - 8 - 5 - 3 - 1(i = 5)$$

$$1 - 2 - 4 - 9 - 5 - 3 - 1(i = 5)$$

$$1 - 2 - 6 - 10 - 12 - 11 - 7 - 3 - 1(i = 7)$$

$$1 - 2 - 6 - 10 - 13 - 11 - 7 - 3 - 1(i = 7)$$

Using equation (3.1), we can now sum up the quantity $(\frac{1}{2})^i$ over all loops, giving $F = (\frac{1}{2})^5 + (\frac{1}{2})^5 + (\frac{1}{2})^7 + (\frac{1}{2})^7 = 0.078125$.

Estimating Inbreeding from Marital Data

The examples given above all compute the inbreeding coefficient by looking at the common ancestry of a child. In many studies of human populations, inbreeding is computed using marital data, where we look at the genealogical relationship of bride and groom and then compute what the inbreeding coefficient of their children would be. In addition to the assumption that there are no extramarital relationships, we also need to assume that the number of offspring is not related to the level of inbreeding when we pool inbreeding coefficients for the entire population.

D. Mean Inbreeding

The discussion of inbreeding so far has concentrated on *individuals* within a population. Not all matings will be the same. Some matings in any human

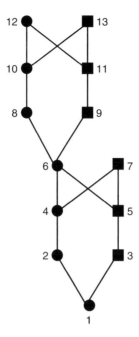

FIGURE 3.11 Example of a genealogy where one of the common ancestors is inbred. Person 1 is inbred because her parents (persons 2 and 3) are first cousins. The common ancestors in this case are persons 6 and 7, who are the grandparents in common for persons 2 and 3. Normally, we would use equation (3.1) to compute an inbreeding coefficient of $F = 0.0625$. However, in this case we see that one of the common ancestors (person 6) is inbred, specifically the product of a first cousin mating between persons 8 and 9. In cases where we have one or more inbred common ancestors, we use equation (3.2), $F = \sum \left(\frac{1}{2}\right)^i (1 + F_A)$, where F_A is the inbreeding coefficient of a common ancestor. For person 1, we have two paths, each of length $i = 5$, tracing back to common ancestors 6 and 7. The inbreeding coefficient of these common ancestors is $F_A = 0.0625$ for person 6 (who is the child of a first cousin mating) and $F_A = 0$ for person 7 (whose parents, not shown, are unrelated). We now solve for equation (3.2) as $F = \sum(\frac{1}{2})^i(1 + F_A) = [(\frac{1}{2})^5(1 + 0.0625)] + [(\frac{1}{2})^5(1 + 0)] = 0.064453125$.

population will be between cousins or closer relatives, but the majority will be between unrelated mates. The frequency of different types of matings varies across human societies, and is influenced by cultural rules and preferences, as well as demographic factors affecting the number of available mates. Some aspects of these factors will be discussed later in this chapter.

From the perspective of population genetics, we need to look at the total amount of inbreeding in a population, characterized by the mean (average) inbreeding coefficient of the population. Consider the following example of a study of inbreeding in the village of Shiiba in Japan. Schull (1972) presents data on 1246 couples; of these, 189 were first cousins, 47 were first cousins once removed, 64 were second cousins, and 946 were unrelated. To calculate the mean (average) inbreeding level in this population, we simply multiply the number of marriages in each category by the respective inbreeding coefficient, sum the results, and divide by the total number of marriages. In mathematical terms, the mean inbreeding

TABLE 3.2 Computation of the Mean Inbreeding Coefficient for the Village of Shiiba, Japan[a]

Type of Marriage	Number of Marriages (n)	Inbreeding Coefficient (F)	Product ($=nF$)
First cousin	189	0.0625	11.8125
First cousin once removed	47	0.03125	1.46875
Second cousin	64	0.015625	1.0
Unrelated	946	0.0	0.0

[a]In order to compute the mean inbreeding coefficient, we first multiply the number of marriages in each category by the respective inbreeding coefficient for that type of relationship. For example, there were $n = 189$ first cousin marriages. The inbreeding coefficient for children born to a first cousin marriage is $F = 0.0625$. The product (nF) is $189(0.0625) = 11.8125$. We repeat this procedure for each type of marriage, and then sum the results: $\sum nF = 11.8125 + 1.46875 + 1.0 + 0.0 = 14.28125$. We then divide this sum by the total number of marriages ($\sum n = 189 + 47 + 64 + 946 = 1246$) to get the mean inbreeding coefficient using equation (3.3):

$$(\sum nF / \sum n) = \frac{14.28125}{1246} = 0.0115.$$

Source: Data from Schull (1972).

coefficient of a population is

$$\frac{\sum nF}{\sum n} \tag{3.3}$$

This example is worked out in detail in Table 3.2. In this case, the mean inbreeding coefficient is $F = 0.0115$, which is close to the value we would obtain if every marriage were between second cousins (0.0156). The mean inbreeding coefficient is what we need to consider when dealing with the genetic impact of inbreeding on the *population*.

II. POPULATION GENETICS AND INBREEDING

Inbreeding increases the relative frequency of homozygotes and decreases the relative frequency of heterozygotes compared with the frequencies expected under Hardy–Weinberg equilibrium. By itself, however, inbreeding does not change allele frequencies.

A. The Impact of Inbreeding on Genotype Frequencies

To show the genetic impact of inbreeding, we start with the standard model of a locus with two alleles, A and a. The frequency of homozygotes (AA, aa) will be due in part to identity by state and in part to identity by descent. It is convenient in this context to think about the inbreeding coefficient F as representing the proportion of the population that is inbred. This means that the proportion of the population that is *not* inbred is $1 - F$. The frequency of genotype AA for the noninbred part of the population is what we expect under Hardy–Weinberg equilibrium, which is p^2. In order to determine the frequency of the AA homozygote that reflects identity by state (where the two A alleles do *not* come from a common ancestor), we are asking about the probability of having the AA genotype in someone in the noninbred part of the population. The probability of having the genotype AA under random mating is p^2. The probability of not being inbred is $1 - F$. To get the probability of having AA and *not* being inbred is the product of these probabilities, which is $p^2(1 - F)$. This gives us the probability of identity by state.

In order to calculate the total frequency of AA genotypes, we also need to know the frequency expected because of identity by descent. Under inbreeding, which by definition means inheriting two copies of the *same* allele from a common ancestor, the only genotypes possible are AA and aa. The probability of having the AA genotype is simply the probability of having the A allele, because an inbred person having a given allele by definition has two copies of that allele. Thus, the frequency of genotype AA for those that are inbred is p (and, by extension, the frequency of genotype aa for those that are inbred is q). If we want to know the frequency of the genotype AA in the population because of identity by descent, we need to multiply the probability of having genotype AA under inbreeding (which is p) by the probability of being inbred (which is F), which gives us pF.

We now can figure out the total frequency of the AA genotype by adding the frequency expected under identity by state [$p^2(1 - F)$] and the frequency expected under identity by descent (pF), which gives $p^2(1 - F) + pF$. This equation can be simplified. First, we multiply out all terms to get $p^2 - p^2F + pF$, which, by factoring pF from the last two terms, gives us $p^2 + pF(1 - p)$, and, as we saw in Chapter 2, $(1 - p) = q$, this means that the expected frequency of genotype AA under inbreeding reduces to $p^2 + pqF$.

The expected frequency of genotype aa could be derived in similar fashion, which gives $q^2 + pqF$. The expected frequency of the heterozygote is easier, because it does not occur under inbreeding, and therefore is simply the probability of not being inbred ($1 - F$) times the probability of being heterozygous ($2pq$), which equals $2pq(1 - F)$.

Putting all of this together gives us the expected genotype frequencies under inbreeding:

$$\text{Expected frequency of genotype } AA = p^2 + pqF$$

$$\text{Expected frequency of genotype } Aa = 2pq(1 - F)$$

$$\text{Expected frequency of genotype } aa = q^2 + pqF \tag{3.4}$$

The different components (noninbred and inbred) of these values are summarized in Table 3.3. There are two things to note from equation (3.4): (1) the sum of all genotype frequencies still adds up to 1.0 (try it); and (2) when there is no inbreeding ($F = 0$), then equation (3.4) reduces to Hardy–Weinberg equilibrium.

As an illustration of the effect that inbreeding has on genotype frequencies, consider the example of a locus with two alleles, A and a, with allele frequencies of $p = 0.8$ and $q = 0.2$, and a mean inbreeding of $F = 0.0625$ (i.e., a population made up entirely of the offspring of first cousin matings). We use equation

TABLE 3.3   Summary of Genotype Frequencies Expected under Inbreeding

Genotype	Homozygotes Identity by State	Identity by Descent	Heterozygotes	Total
AA	$(1-F)p^2$	pF	—	$(1-F)p^2 + pF = p^2 + pqF$
Aa	—	—	$2pq(1-F)$	$2pq(1-F)$
aa	$(1-F)q^2$	qF	—	$(1-F)p^2 + pF = p^2 + pqF$

(3.4) to compute these frequencies and compare them to those expected under Hardy–Weinberg equilibrium:

Genotype	$F = 0.0625$	Hardy–Weinberg $(F = 0)$
AA	0.65	0.64
Aa	0.30	0.32
aa	0.05	0.04

At this level of inbreeding, the changes in the genotype frequencies are relatively minor but do show the increase in frequency of homozygotes and the decrease in frequency of heterozygotes relative to Hardy–Weinberg equilibrium.

B.   Why Inbreeding Does Not Change Allele Frequencies

Inbreeding is not considered an evolutionary force because it does not directly cause a change in allele frequencies. This is easy to see if we use equation (2.3) to compute the allele frequency in the next generation (p') from the genotype frequencies expected under inbreeding [which we get from equation (3.4)], giving

$$p' = f_{AA} + \frac{f_{Aa}}{2}$$
$$= (p^2 + pqF) + \frac{2pq(1-F)}{2}$$

which reduces to

$$p' = (p^2 + pqF) + pq(1-F)$$
$$= p^2 + pqF + pq - pqF$$
$$= p^2 + pq$$
$$= p(p+q)$$

Finally, because $(p+q) = 1$, we get $p' = p$; that is, the allele frequency does not change. By itself, inbreeding does not cause evolution.

C.   The Medical Impact of Inbreeding

Inbreeding becomes significant in evolution because it can affect the rate of natural selection. Consider that many genetic diseases stem from harmful recessive alleles. In order to manifest such conditions, a person needs to be homozygous for the recessive allele, which means inheritance of the recessive allele from both parents. Because inbreeding increases the probability of homozygosity, populations with higher levels of inbreeding will have a greater proportion of homozygotes to be affected by natural selection. Examples of the interaction of inbreeding and natural selection will be given in a later chapter. For now, we will focus on a short review of the medical impact of inbreeding.

If a recessive allele is rare (of low frequency), then the expected frequency of recessive homozygotes under Hardy–Weinberg equilibrium is very rare (q^2). Under inbreeding, this proportion can increase quite a bit, such that rare genetic disorders become more prevalent. For example, if a recessive allele exists at frequency $q = 0.01$, then the frequency of recessive homozygotes under Hardy–Weinberg equilibrium is $q^2 = 0.0001$. Under first cousin mating ($F = 0.0625$), from equation (3.4) we obtain the quantity $q^2 + Fpq = 0.00072$, which is over seven times as many recessive homozygotes.

Numerous studies have found an association between **consanguineous marriages** (marriage between kin) and a variety of health conditions, including congenital defects and mental retardation (Bittles 2001), as well as links with some complex diseases (Bittles and Black 2010). On the other hand, there is less evidence for an effect of inbreeding on fetal loss (Bittles 2001). Overall, there is a clear link between inbreeding and prereproductive mortality (dying before adulthood). In a comparative analysis of inbreeding and prereproductive mortality in 69 human populations in four continents, Bittles and Black (2010) found an average of 3.5% excess mortality among children born to first cousins relative to unrelated couples. Because of the complex interplay between cultural, demographic, and genetic factors that affect health, more work needs to be done to understand better the role of inbreeding on human fertility, mortality, and morbidity.

III.   INBREEDING IN HUMAN POPULATIONS

In addition to the more general questions about the genetic impact of inbreeding, anthropologists are interested how inbreeding affects genetic diversity in human populations as well as the effect of culture on levels of inbreeding. The remainder of this chapter examines variation in inbreeding in human populations, including several case studies that illustrate different ways of assessing inbreeding in human societies, including one method that is unique to human genetics.

A.   Rates of Inbreeding in Human Populations

Total inbreeding levels in humans tend to range from close to zero to values of roughly 0.05 (Reid 1973; Bittles 2001). Although low relative to some other species, there is considerable variation both within and among human populations in terms of the frequency of consanguineous marriage and inbreeding. This is not surprising, given the role of culture and demography on shaping human mate choice. Some cultures have restrictions on consanguinity, whereas others have preferences for certain types of consanguineous marriage. Demographically, small populations (typical of much of human evolution) pose a problem in that one can quickly run out of potential mates within a local population that are not closely related.

Of course, variation in the rate of consanguineous matings depends on the level of relatedness. It is important here to differentiate between *inbreeding* as a general term and the more restricted term *incest*. **Incest** is defined here as the mating between relatives that is closer than generally permitted in the society, which for many human groups tends to be associated with a first-degree genetic relationship; that is, between siblings or between parent and child. Such matings are almost universally taboo, although there have been exceptions historically, such as cases in some ancient Egyptian dynasties (Bittles 2004).

Apart from first-degree (incestuous) relationships, human cultures vary considerably in their avoidance of or preference for other forms of inbreeding. Consequently, there is considerable variation in the prevalence of consanguineous marriages across the world. Bittles (2001) examined global variation in close consanguineous marriages, defined as marriage between second cousins or closer. He found that rates of consanguineous marriage tend to be low in North America, Australia, and most of Europe. Parts of the Middle East, South and Central Asia, and North Africa include many populations where between 20% and >50% of all marriages are consanguineous.

Much of this variation is associated with religion. The Roman Catholic Church prohibits marriages of first cousins or closer, and until 1917 had prohibited marriages between second and third cousins. Various Protestant denominations have no restriction on first cousin marriage, and neither do Buddhists. Some Muslim populations are characterized by a preference for certain types of first cousin marriage, such as a marriage between a man and his father's brother's daughter, as compared with his mother's brother's daughter (Bittles 2004; Bittles and Black 2010). Such examples show that cultural rules regarding inbreeding are not always based on genetic principles; these two relationships are the same genetically.

There is also variation in consanguineous marriages within societies, often seen in differences in civil legislation associated with marriage. The United States provides a good example of this type of variation when considering laws against first cousin marriage. Although first cousin marriage is permitted in 20 states, it is illegal in 22 other states, and a criminal offense in eight other states. There are also variations within states. In Wisconsin, for example, first cousin marriages are permitted if one of the individuals is sterile or if the woman is over 55 years of age, and thus considered likely to be sterile (Bittles 2010).

B. Examples of Inbreeding Studies Using Genealogical Data

Two brief case studies are presented here to give examples of how genealogical data are used to compute and interpret inbreeding in human populations.

The Romany of Wales

The Romany (often referred to as "gypsies") are a European ethnic group with original roots in India. Williams and Harper (1977) studied a Romany group living in South Wales, most of who were descended from the marriage of a Romany and non-Romany Englishman early in the nineteenth century. This population was of interest because of the high prevalence of phenylketonuria (a genetic disorder characterized by an enzyme deficiency) and other recessive disorders. Analysis of 99 matings found 14 between first cousins and 22 between second cousins, and the remaining 63 were unrelated. Using the inbreeding coefficients for first and second cousins from Table 3.1 and equation (3.3), we can estimate the mean inbreeding coefficient for the population as

$$\frac{\sum nF}{\sum n} = \frac{14(0.0625) + 22(0.015625) + 63(0)}{99} = 0.012$$

This is a minimum estimate that assumes that the noncousin matings were completely random ($F = 0$). Williams and Harper (1977) note that this is not likely the case and adjust this figure upward to account for consanguinity among the other matings between Romany, obtaining a value of $F = 0.017$. This is a relatively high level of inbreeding compared with larger European populations, consistent with some degree of **endogamy** (the tendency to choose mates within a group, as opposed to **exogamy**, the tendency to choose mates outside the group). Williams and Harper note that this group is not completely endogamous; 24 out of 99 matings were exogamous, between a Romany and a non-Romany.

A problem with a number of studies, such as the Welsh gypsy study just mentioned, is that they essentially reflect a single point in time rather than the ongoing evolution of a population. In many cases we are left unable to see how "current" estimates of inbreeding and exogamy fit into the overall history of a population. Are such rates stable, or do they change over time? If they change, are such changes essentially random fluctuations, or are they indicative of a trend? If a trend is present, then what is its cause? To answer such questions, we need data from more than one point in time.

The Ramah Navajo

Use of large genealogies can provide information about the impact on inbreeding resulting from matings of remote ancestors even when close relatives, such as first cousins, are avoided. Spuhler and Kluckhohn (1953) collected genealogies on the Ramah Navajo, a group of several hundred Navajo Indians living near Ramah in the westcentral part of New Mexico. The Ramah Navajo population began about 1869 when the US Army released captive Navajo. Navajo culture prohibits marriage within one's own clan or one's father's clan, or with clans that are linked to one's clan or one's father's clan. These beliefs lead to clan exogamy, which tends to reduce levels of inbreeding. In addition, first cousin marriages are forbidden (although four were observed in the data). There was, however, a number of other more remote forms of inbreeding that were detected from the extensive time depth of the genealogies, ranging from first cousin once removed to half fourth cousins. The initial analysis (Spuhler and Kluckhohn 1953) showed a low level of total inbreeding of $F = 0.0066$. Closer analysis showed that a large amount of the total inbreeding coefficient resulted from the more remote forms of consanguinity, such as second, third, and fourth cousins. Additional research including the time period between 1950 and 1964 increased the estimate of overall inbreeding to $F = 0.0092$ (Spuhler 1989). This value is still relatively low, showing the effect of avoidance of close relatives as spouses, but does show how inbreeding can be underestimated if we lack sufficient historical depth.

C.   Surname Analysis

Studying inbreeding in human populations is facilitated greatly by the fact that humans keep written records and other organisms do not! Another behavior of humans that helps us in studying inbreeding is the tendency in many cultures to inherit one's surname (last name) from one's father. Like the Y chromosome, a person's surname passes on from father to son in many societies. In 1875, George Darwin, the son of Charles Darwin, attempted to estimate the frequency of cousin marriages on the basis of marriages between people with the same surname (Lasker 1985). A more complete analysis and approach was later developed by Crow and Mange (1965) which allowed estimation of the mean inbreeding coefficient of a population from marriage records by noting how frequently the bride and groom had the same last name. The term **isonymy** is used to denote marriages where the surnames of bride and groom are the same (and where the bride's surname is her maiden name).

Computing F From Marital Isonymy

Crow and Mange noted that for many different forms of consanguinity, the expected frequency of isonymy was 4 times that of the inbreeding coefficient of two parents. For example, if a bride and groom are first cousins, they share a set of grandparents. As shown in Figure 3.12, there are four different ways that first cousins can be related:

1. When the groom's father and the bride's father are siblings (brothers). Here, the groom and the bride will have the same last name.
2. When the groom's father and bride's mother are siblings (brother and sister).
3. When the groom's mother and bride's father are siblings (sister and brother).
4. When the groom's mother and the bride's mother are siblings (sisters).

Because surnames are passed through the male line, only one-fourth of the first cousin marriages will be isonymous—those cases where the groom's father and the bride's father are brothers (Figure 3.12a).

Crow and Mange (1965) considered other types of relationships. For example, among second cousin marriage, the probability of isonymy is 1 in 16. Their great insight came from the realization that the expected frequency of isonymy was directly proportional to the expected inbreeding coefficient for each type of relationship. For first cousin marriage, the expected frequency of isonymy is $\frac{1}{4}$

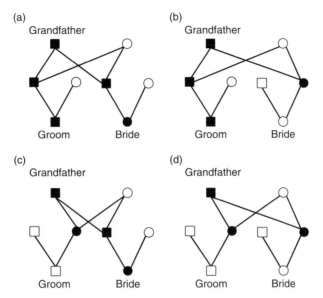

FIGURE 3.12 Inheritance of surnames in first cousin marriages. First cousins share a set of grandparents. There are four possible ways for the parents of first cousins to be related: (a) the groom's father and the bride's father are brothers, (b) the groom's father and the bride's mother are brother and sister, (c) the groom's mother and the bride's father are sister and brother, and (d) the groom's mother and the bride's mother are sisters. Each diagram shows the inheritance of the grandfather's surname—individuals inheriting this surname are shown by filled squares (males) and circles (females). Males inherit their father's surname and pass it on to their children. Females inherit their father's surname (as their maiden name) but do not pass it on. Only diagram (a) depicts isonymy, where both groom and bride have the same surname. Among possible first cousin marriages, only one of four will be isonymous.

and the inbreeding coefficient of their children (see Table 3.1) is $\frac{1}{16}$. The ratio of isonymy to inbreeding is $\frac{1}{4}/\frac{1}{16} = 4$. For second cousins, the expected frequency of isonymy is $\frac{1}{16}$ and the inbreeding coefficient of their children is $\frac{1}{64}$, giving the same ratio: $\frac{1}{16}/\frac{1}{64} = 4$. Crow and Mange (1965) found the same ratio for a number of different degrees of relationship. They astutely reasoned that if we know the overall frequency of isonymy in a population, we could estimate the mean inbreeding level for the population by using this ratio and a very simple formula

$$F = \frac{P}{4} \tag{3.5}$$

where P is the relative frequency of isonymous marriages in a population. (Another method of estimating F from isonymy is shown in the next section.)

The isonymy method is not perfect. For one thing, in some pedigrees the ratio of isonymy to inbreeding is not equal to 4 (Crow and Mange 1965; Crow 1980). Isonymy can sometimes underestimate F when common ancestors are inbred. On the other hand, isonymy can sometimes overestimate F when there are multiple origins of the same surname. Some surnames, such as Smith or Taylor, derive from occupation, and can be common in a population but not necessarily indicate common ancestry. Another problem is the time depth of the estimates. Any inbreeding estimate depends on a reference generation. With genealogical data, this reference population is the oldest generation for which data are available. With surname data, the reference population is a hypothetical founding group where each male has a unique surname (A. Rogers 1991). Consequently, inbreeding estimates from isonymy tend to be larger than those that are estimated from pedigree data. Other potential problems include changes in spelling, adoptions, and illegitimate births. Despite problems, there is agreement between isonymy estimates and pedigree estimates in some cases depending on the composition of the samples (L. Rogers 1987). Further, while the absolute estimates of F may be overestimated, comparison of relative values, such as between populations or time periods, can still be valuable.

Random and Nonrandom Components of Inbreeding

If we have an estimate of the level of inbreeding from a population, then what exactly does this tell us about mating behavior? Does a relatively high value, such as $F = 0.05$, tell us that there is a preference for consanguineous mating or marriage in the population? Not necessarily. In small populations, a certain amount of inbreeding might occur simply by chance because most individuals in a small population might be related to begin with. If every potential mate is a cousin, then even selecting a mate at random will result in inbreeding. Crow and Mange (1965) showed how to partition these effects using surname data from marriage records, and this method was later expanded upon by Crow (1980). They showed that the *total* amount of inbreeding in a population (F) is made up of a random inbreeding component (F_r) and a nonrandom inbreeding component (F_n). Random inbreeding is the amount of inbreeding expected by chance because of finite population size (i.e., the smaller the population, the fewer mates available that are not related to you). Nonrandom inbreeding is the amount of inbreeding that is due to the net effect of avoidances and preferences for consanguineous marriages.

The computation of these components is shown here using the Hutterite data from Crow and Mange (1965). In order to compute these components, we need the observed probability of isonymous marriages P and the expected probability of isonymous marriages under random mating P_r. Among the Hutterites, 87 in 446 marriages were isonymous, giving $P = \frac{87}{446} = 0.195$. We then compute the probability of random martial isonymy by pairing each male in the sample with each female in the sample and counting the proportion of those pairs that have the same last name. This sounds complicated, but is easy to do from a table of surnames, as shown in Example 3.1, which gives random marital isonymy of $P_r = 0.178$. The random and nonrandom components of inbreeding (F_r and F_n, respectively) are then computed as

$$F_r = \frac{P_r}{4}$$

$$F_n = \frac{P - P_r}{4(1 - P_r)} \tag{3.6}$$

and total inbreeding now becomes

$$F = F_n + (1 - F_n)F_r \tag{3.7}$$

Total inbreeding in equation (3.7) is not simply the sum of nonrandom and random components; since inbreeding can be either nonrandom *or* random, the random component must be multiplied by the probability of nonrandom inbreeding $(1 - F_n)$.

EXAMPLE 3.1 Computing the Probability of Random Marital Isonymy. We can compute the amount of random marital isonymy (marriages where the bride and groom have the same surname) by pairing each male with each female and then counting the number of times that they have the same last name. An easier way to do this is to construct a table of all surnames in the population and count the number of times males and females have a given surname.

The table below shows data on surnames from 446 marriage records of the Hutterites as reported by Crow and Mange (1965). The column labeled m is the proportion of males with a given abbreviated surname. For example, 7 of 446 males have the surname "De," giving the proportion $m = \frac{7}{446} = 0.016$. The column labeled f is the proportion for the females. The last column (mf) is the product of m and f (e.g., 0.016 times 0.020 = 0.00032).

Surname	Males	Females	m	f	mf
De	7	9	0.016	0.020	0.000320
Gl	8	7	0.018	0.016	0.000288
Gr	36	27	0.081	0.061	0.004941
Ho	104	93	0.233	0.209	0.048697
Kl	25	27	0.056	0.061	0.003416
Ma	25	19	0.056	0.043	0.002408
St	17	15	0.038	0.034	0.001292
Ts	7	8	0.016	0.018	0.000288
Wd	134	153	0.300	0.343	0.102900
Wi	30	33	0.067	0.074	0.004958
Wo	41	38	0.092	0.085	0.007820
Wu	12	17	0.027	0.038	0.001026
Total	446	446	1.000	1.000	0.178

We now compute the total probability of random marital isonymy by summing all the values in the last column, giving $P_r = 0.178$. We divide this number by 4 to get the random component of inbreeding (F_r), which is the amount of inbreeding we expect even when there is no preference or avoidance of consanguineous marriage. In this case, $F_r = \frac{0.178}{4} = 0.0445$.

Although the random component of inbreeding will always be positive, the nonrandom component can be positive or negative, depending on the balance between a preference for consanguineous marriage (making the component positive) and avoidance of consanguineous marriage (making the component negative). For the Hutterite data, $P = 0.195$ and $P_r = 0.178$, giving a random inbreeding component of $F_r = 0.0445$, a nonrandom inbreeding component of $F_n = 0.0052$, and a total inbreeding value of $F = 0.0494$. This last value is higher than what we would get using equation (3.5) because there are actually two different methods for estimating F from surnames—see Crow (1980) for a discussion of which method should be used in different circumstances.

What do these numbers mean? The overall level of inbreeding ($F = 0.0494$) is fairly high, which is consistent with the isolated and endogamous nature of the Hutterite population. Most of this quantity is due to the high level of random inbreeding ($F_r = 0.0445$), whereas the nonrandom component of inbreeding is close to zero. These values show that high inbreeding in the Hutterite population is *not* the result of any preference for consanguineous marriage, but instead is a reflection of the small and isolated nature of the population, resulting in a situation where most available mates are related.

Analysis and interpretation of inbreeding components can be complicated and require careful attention to the demographic structure of a population. For example, Devor et al. (1983) looked at marriage records from Ramea Island in Newfoundland, Canada, a population that had been a small and isolated fishing port until the early 1950s, after which time the building of a fish processing plant led to large levels of immigration. Using data from 201 marriages, they found a small random component ($F_r = 0.004$) but large levels of nonrandom ($F_n = 0.0253$) and total ($F = 0.0315$) inbreeding. At first glance, these results suggest that the level of inbreeding in this population was due primarily to a preference for consanguineous marriages because the nonrandom component is large relative to total inbreeding and is positive. However they found that this was not the case. Here, the large nonrandom component was due in large part to the immigration of a large number of isonymous marriages from small populations outside the island. As such, these results tell us more about mating behavior of populations other than that of Ramea Island.

Because isonymy methods can be used with historical data, we have the chance to track changes in inbreeding components over time. Relethford and Jaquish (1988) examined historical marriage records in five towns in northcentral Massachusetts. Figure 3.13 shows the trends in inbreeding components averaged over these towns for five decades from 1800 to 1849. This time period was characterized by rapid population growth and increased migration. All inbreeding components declined over time, reflecting the increase in population size and migration rates, both of which act to reduce the proportion of consanguineous marriages. Random inbreeding accounts for a greater amount of total inbreeding than nonrandom inbreeding. Still, the values of nonrandom inbreeding are positive, which might suggest some slight preference for consanguineous marriage. As with the Devor group's study of Ramea Island, the situation is actually more complex and is affected greatly by a high rate of isonymous exogamous marriages (where one or both spouses was a migrant into the community). Cousin marriage was not prohibited, and there is some suggestion that there was a preference for long-distance consanguineous marriage. Once the isonymous exogamous marriages are removed from the analysis (see Figure 3.14), random inbreeding

FIGURE 3.13 Components of inbreeding for five towns in westcentral Massachusetts, 1800–1849. [*Source of data*: Relethford and Jaquish (1988).]

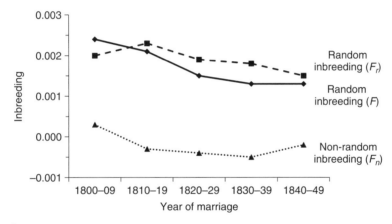

FIGURE 3.14 Components of inbreeding for five towns in westcentral Massachusetts, 1800–1849, excluding isonymous exogamous marriages. [*Source of data*: Relethford and Jaquish (1988).]

accounts for most of the total inbreeding, offset to some extent by an *avoidance* of consanguineous marriages within the populations (see the *negative* values for nonrandom inbreeding in Figure 3.14).

D. Potential-Mates Analysis

A more precise way of looking at random and nonrandom components of inbreeding is provided by **potential-mates analysis**, a set of methods that includes using computer simulation to define characteristics of a pool of potential mates based on pedigree and demographic data (Leslie 1985). Individuals can be paired up at random with realistically defined potential mates, based on sex, age, and

other relevant characteristics. In the case of inbreeding, the inbreeding coefficients of randomly chosen pairs of potential mates can be computed (the random
component) and then compared with the observed inbreeding coefficient of the
actual mates.

Brennan and Relethford (1983) used this method to look at random and nonrandom inbreeding coefficients for the island of Sanday, located in the northern
part of the Orkney Islands, Scotland. Pedigree data were used to compute inbreeding based on marital pairs for three time periods according to the year of birth
of the husband: 1855–1884, 1885–1924, and 1924–1964. Random inbreeding was
computed from all possible pairings of potential mates that met certain realistic
criteria for mating, such as the age of each potential spouse, the age difference
between the potential husband and wife, and excluding anyone from the same
nuclear family (no incest). Given the observed inbreeding coefficient and the
random inbreeding coefficient, Brennan and Relethford computed the nonrandom
component using equation (3.7).

The results are shown in Table 3.4. Over time, total observed inbreeding
declined, reaching zero in the last time period. Random inbreeding decreased
and then increased over time, while nonrandom inbreeding declined from a slight
amount of consanguinity preference (a positive nonrandom component) to consanguinity avoidance (a negative nonrandom component in the last time period). The
change in nonrandom inbreeding is related to cousin marriage because the choice
of potential mates ruled out closer, incestuous marriages. Brennan and Relethford
(1983) were able to relate these shifts in mating behavior to demographic trends
since the midnineteenth century. Over time, random inbreeding was offset by a
change in nonrandom inbreeding to a pattern of avoidance of cousin marriage; such
avoidance was made possible by an increase in the number of individuals seeking a
mate off-island as well as an increase in the geographic distance between on-island
spouses. A small number of local potential mates was thus countered by increased
mobility, allowing avoidance of cousin marriages, and resulting in a net inbreeding
value of zero in the last time period. Potential-mates analysis allows a more precise analysis and interpretation of the complex interplay between population size,
migration, cultural preferences, and inbreeding than can be obtained from surname
analysis.

**TABLE 3.4 Components of Inbreeding from Potential-Mates Analysis of
Sanday, Orkney Islands, Scotland**

Birth Year of Husband	Total Inbreeding	Random Inbreeding	Nonrandom Inbreeding
1855–1884	0.00212	0.00120	0.00092
1885–1924	0.00091	0.00074	0.00017
1925–1964	0.00000	0.00083	−0.00083

Source: Data from Brennan and Relethford (1983).

IV. SUMMARY

Inbreeding is the mating of genetically related individuals. Simple mathematics shows that we are all inbred to some extent, but in practice, the genetic and medical significance of inbreeding is for individuals who are cousins or closer. The inbreeding coefficient is a measure of the probability that an individual will have identical alleles due to common ancestry (identity by descent). Inbreeding coefficients are computed from genealogical information by counting the number of ancestors along a path back to a common ancestor of both parents. First cousins, for example, have a set of grandparents in common, whereas second cousins have a set of great grandparents in common. On a populational level, mean inbreeding can be derived as the average inbreeding coefficient over all matings.

Inbreeding increases the frequency of homozygotes in a population and decreases the frequency of heterozygotes. Although inbreeding changes genotype frequencies from those expected under Hardy–Weinberg equilibrium (which assumes random mating, or $F = 0$), it does *not* change allele frequencies. Because inbreeding increases homozygosity, this means that inbred individuals have a greater chance of inheriting two copies of recessive alleles, which in many cases can have significant medical consequences.

Mean inbreeding coefficients for human populations would be less than expected if the entire population were made up of first cousin matings. Although all human societies prohibit incest (extremely close inbreeding from parent–child or sib–sib (intersibling) mating), there is a great deal of cultural variation regarding cousin marriage. Some cultures discourage such marriages/matings, whereas other cultures have a preference for cousin marriage. Because of the way in which surnames are inherited in many societies, the frequency of marriages between people with the same last name can give us an estimate of inbreeding.

Surname frequencies can also be used to partition the total level of inbreeding into two components—random inbreeding, reflecting the marriages between relatives expected by chance due to small population size, and nonrandom inbreeding, which gives us an idea as to the net effect of avoidance of, or preference for, mating between relatives.

CHAPTER 4

MUTATION

Where do new genetic variants come from? What is the origin of variation? Darwin recognized that variation existed, and that natural selection acted on this variation, but the molecular nature of genetic change was unknown in his time. We now know that mutations, changes in the genetic code, occur every generation. Without mutation, there would be no evolution (and, consequently, no life). Mutation (and mutants) is an important cornerstone in much of science fiction, including comic books, such as the *X-Men*. Here, the focus on mutants is on their unique abilities that set them apart from "normal" humans. A recurring theme in much of this fiction is the evolutionary and social divide between mutants and normal people, often portrayed in terms of the oppression and persecution of the mutants. While the use of mutant status as a surrogate for social forms of discrimination is a useful dramatic foil, in reality there would be no such divisions within our species into mutants and "normals." From a statistical viewpoint, we each carry mutant alleles, and are all mutants. In the real world, it is important to remember that mutation is not an unusual phenomenon, but something that occurs all of the time and is vital to the evolution of life.

I. THE NATURE OF MUTATIONS

Mutations are random changes in genetic code. Returning to the idea that genetic transmission is the transmission of information, from cell to cell and from generation to generation, a mutation is an error in the transmission. Any transmission of information is subject to errors, whether this is static on a phone line or a smudged fax transmission. I recall from grade school a game we played one day that involved the transmission of information. The teacher started the game by whispering a message to the first student in the first row of the classroom. This student then whispered the message to the student immediately behind in the second row, who in turn whispered the message to the next student in line. This whispering sequence continued up and down the rows of the class until the last student gave the message to the teacher, who then wrote both the original and last messages on the board. I cannot recall what the message was, but I do remember that it had changed somewhat because of accumulated errors in transmission from

Human Population Genetics, First Edition. John H. Relethford.
© 2012 Wiley-Blackwell. Published 2012 by John Wiley & Sons, Inc.

one student to the next. As a broad analogy, this was an example of a *mutation*. In genetics, the same thing can happen when information is not replicated exactly from one cell to the next. In our body cells, this can cause damage, including cancer. If such a mutation occurs in a sex cell, then this information becomes evolutionarily significant because it changes the transmission of genetic information from one generation to the next.

A. Types of Mutation

Mutations are random changes in the genetic code that occur for a number of reasons, including chemical exposure, ultraviolet radiation, viral infection, and background radiation. Mutations can occur in a variety of ways, some of which are reviewed here briefly.

Point Mutations

The simplest form of mutation is a **point mutation**, where there is a change from one DNA base to another, such as from C to T, or from A to C. The four nucleotide bases have different biochemical structures. Bases A and G are purines, which are defined as having two carbon–nitrogen rings. Bases C and T are pyrimidines, which are defined as possessing one carbon–nitrogen ring. When the mutation is between two purines (A and G) or between two pyrimidines (C and T), it is called a **transition**. When the mutation is between a purine and a pyrimidine (A and C, A and T, G and C, or G and T), it is called a **transversion**. Transitions are much more common than transversions.

Point mutations can have differing effects. If a point mutation occurs in a noncoding part of the DNA, with no impact on the organism, then this mutation can be considered neutral. For coding DNA, some point mutations can be neutral if the mutation results in the same amino acid, which is known as a **silent mutation** (see Table 4.1). An example is the genetic code for the amino acid glycine. There are four different DNA sequences that code for glycine: CCA, CCT, CCC, and CCG. If there is a DNA sequence of CCA and the last base pair changes from A to G, giving the sequence CCG, the code is still for the same amino acid. Note that we would be able to detect this mutation from a direct assessment of the DNA sequence but not from analysis of amino acids—the change would be "silent" at this level. For other point mutations, there can be a more noticeable impact. For example, a point mutation for the sixth amino acid in the beta chain of the hemoglobin molecule results in a change from the amino acid glutamic acid to the amino acid valine, which results in a variant form of hemoglobin known as *sickle cell hemoglobin*. Individuals with two copies of this allele have the genetic disease known as *sickle cell anemia* (see Chapter 7 for more information on the evolution of the sickle cell allele).

Insertions and Deletions

In addition to changes due to substitution in the DNA, mutations can also produce **insertions** and **deletions** of DNA sequences. The term **indel** is used to refer to insertions and deletions collectively. A change in DNA sequence through the

TABLE 4.1 DNA Triplets for Coding of Amino Acids[a]

AAA	phenylalanine	AGA	serine	ATA	tyrosine	ACA	cysteine
AAG	phenylalanine	AGG	serine	ATG	tyrosine	ACG	cysteine
AAT	leucine	AGT	serine	ATT	stop	ACT	stop
AAC	leucine	AGC	serine	ATC	stop	ACC	tryptophan
GAA	leucine	GGA	proline	GTA	histidine	GCA	arginine
GAG	leucine	GGG	proline	GTG	histidine	GCG	arginine
GAT	leucine	GGT	proline	GTT	glutamine	GCT	arginine
GAC	leucine	GGC	proline	GTC	glutamine	GCC	arginine
TAA	isoleucine	TGA	threonine	TTA	asparagine	TCA	serine
TAG	isoleucine	TGG	threonine	TTG	asparagine	TCG	serine
TAT	isoleucine	TGT	threonine	TTT	lysine	TCT	arginine
TAC	methionine	TGC	threonine	TTC	lysine	TCC	arginine
CAA	valine	CGA	alanine	CTA	aspartic acid	CCA	glycine
CAG	valine	CGG	alanine	CTG	aspartic acid	CCG	glycine
CAT	valine	CGT	alanine	CTT	glutamic acid	CCT	glycine
CAC	valine	CGC	alanine	CTC	glutamic acid	CCC	glycine

[a]Triplets of DNA base pairs allow for the coding of 20 amino acids and a termination sequence (stop). Because there are 64 possible DNA sequences, this means that some sequences code for the same amino acid, allowing for silent mutations, such as a mutation from A to T in the triplet CCA, giving CCT. Both CCA and CCT code for glycine, so this would be a silent mutation. Note that this table provides the codes for the DNA triplets—to convert to messenger RNA codons, substitute U (uracil) for A, A for T, G for C, and C for G.

insertion or deletion of base pairs can change the function of a gene. An example is found in the CCR5 gene in humans, found on chromosome 21, which codes for a receptor protein. One mutant allele is known as the *CCR5-Δ32* allele (Δ is the Greek letter delta, often used to denote difference). This mutation consists of a 32-bp deletion in the gene, which changes the biochemical properties of the CCR5 protein. One interesting consequence of this allele is that people who inherit two copies (having the *CCR5-Δ32–CCR5-Δ32* genotype) are resistant to infection from HIV, the virus that causes AIDS (more information on this mutation is given in Chapter 7).

Chromosomal Changes

Mutations can also affect large sections of entire chromosomes. Such changes include **inversions**, where a section of a chromosome winds up in reverse order. Mutation can also cause deletion of large chunks of a chromosome. **Translocations** occur when sections of a chromosome move to another chromosome; some of these involve an exchange of DNA between chromosomes, and some do not. Entire chromosomes can also be replicated. Down syndrome, for example, is caused when there is an extra twenty-first chromosome.

B. The Evolutionary Impact of Mutation

Three major points need to be kept in mind regarding mutation and evolution. First, mutation is the ultimate source of all genetic variations. Whether this is diversity in blood types, single nucleotide polymorphisms (SNPs), or other genetic traits, different alleles start out as mutations. The second major point, as noted earlier, is that a mutation will have a direct evolutionary consequence only if it occurs in a sex cell, because that is the only way that a mutation can be passed on to the next generation. The third point to keep in mind is that mutation is random with respect to its evolutionary significance. In other words mutations do not appear because they would be useful for a population to possess at a given point in time.

Mutation is an evolutionary force because it leads to a change in allele frequency over time. To illustrate this, imagine a locus with a single allele A in a population of 2000 people. Because all 2000 people have two copies of the A allele (genotype AA), there are 4000 A alleles in the population. Now, imagine that a mutation occurs in one of the A alleles, leading to a new allele designated as a. We now have 3999 A alleles and one a allele. The frequency of the A allele has changed from $p = \frac{4000}{4000} = 1.0$ to $\frac{3999}{4000} = 0.99975$. The frequency of the a allele has changed from $q = \frac{0}{4000}$ to $\frac{1}{4000} = 0.00025$. Admittedly, this is a very small change, but it is still a change. Remember that one of the assumptions of Hardy–Weinberg equilibrium is that there are no mutations. When mutations do occur, the allele frequencies change from those expected under Hardy–Weinberg equilibrium.

The major evolutionary significance of mutation is to introduce new alleles into a population. By itself, mutation does not lead to major shifts in allele frequency. As such, mutation is a *necessary, but not sufficient*, cause of evolution. Mutation is the ultimate source of all genetic variation, and evolution would not be possible without it. On the other hand, evolution is not simply the accumulation of mutations. By itself, mutation can only introduce new genetic variants into a population, but cannot lead to major changes in allele frequencies over reasonable lengths of evolutionary time. The other evolutionary forces (selection, drift, gene flow) act on new mutants as they arise in a population, causing them to sometimes decrease, and sometimes increase, in frequency. Thus, discussion of some of the population genetics of mutation is actually deferred until later in this book when interactions of mutation with other evolutionary forces are considered.

C. Rates of Mutation

Mutations can be regarded as relatively rare events when we consider the probability of a mutation occurring in an individual at a given locus in a given generation. Although generally low at this level, the exact rate of mutation varies across different parts of the genome and among species. Rates also vary depending on whether we are talking about a specific base pair, an entire locus, or the entire genome. For example, in humans the average mutation rate for a single site is about 2.3×10^{-8} (0.000000023) per base pair per generation. Of these, transitions tend to occur more frequently than transversions, and some transitions have even higher rates (''hotspots'' of mutation) (Jobling et al. 2004). Indel mutations have lower mutation rates, averaging about 2.3×10^{-9}. Repeated units of DNA, such as short tandem repeats and minisatellites, have higher mutation rates (Rosenberg et al. 2003b; Jobling et al. 2004). Mitochondrial DNA has higher mutation rates than does nuclear DNA (Stoneking 1993). Regardless of this variation, the main point to remember is that even for rapidly mutating sections of our genome, mutation alone cannot lead to major changes in allele frequency, even over long periods of time.

II.   MODELS OF MUTATION

For the moment, several basic models of mutational change are shown in order to give the reader a better feel for the evolutionary impact of mutation and serve as a baseline for later consideration of other evolutionary forces. More complex models that deal with the interaction of mutation with other evolutionary forces will be presented in later chapters.

A.   A Simple Mutation Model

We start with a very simple mutation model of a single locus starting with one allele, (A) and allowing for the probability of an A allele mutating into another allele (a) with each generation. We assume that this is the only form of mutation (in other words, that there is only one mutant allele) and that mutation is irreversible (i.e., that an a allele cannot mutate back into an A allele). Other than that, we keep all of the other assumptions of Hardy–Weinberg equilibrium—no genetic drift, no gene flow, and no selection for or against the mutant allele. The key parameter here is the mutation rate, represented as μ, the Greek letter mu. Mutation is random, so this rate represents the *probability* of mutation per locus per generation. The actual number of mutations in any generation might be higher or lower, and the mutation rate is an estimate of the average probability of mutation over long periods of time.

Now that we are considering change in allele frequencies from one generation to the next, we use allele frequencies with subscripts indicating the specific generation. Here, p_t refers to the frequency of allele A in generation t, and q_t refers to the frequency of allele a in generation t. As an example, the expression p_3 refers to allele frequency p in generation 3.

When using subscripts, we generally start at $t = 0$. Here, let us start with allele frequencies p_0 and q_0 for the initial frequencies of the A and a alleles. Under Hardy–Weinberg equilibrium, the next generation will have the same frequencies, and so on into the future. Under our simple mutation model, things operate a bit differently. We are using a model where some A alleles mutate into a alleles with a probability of μ. In the next generation, some A alleles will have mutated but most will remain A alleles. Given that μ is the probability of A mutating into a, then $(1 - \mu)$ is the probability of an A allele *not* mutating but instead remaining an A allele. We now ask the question: What is the expected frequency of the A allele in generation $t = 1$? Another way of expressing this is to ask the probability of an allele being A in the previous generation (which is p_0) *and* not mutating (which is $1 - \mu$). The answer is simply the product of the two probabilities, which gives

$$p_1 = p_0(1 - \mu) \tag{4.1}$$

If, for example, we start with $p_0 = 1.0$ with a mutation rate of $\mu = 1 \times 10^{-6} = 0.000001$, then $p_1 = 0.999999$.

We can now go further and consider what will happen in generation 2 after an additional generation of mutation. Following the same logic as used to derive equation (4.1), we can express the expected frequency of the A allele in generation 2 as a function of the expected frequency of the A allele in generation 1 and the mutation rate

$$p_2 = p_1(1 - \mu) \tag{4.2}$$

which would give $p_2 = 0.999998$. We could continue performing this calculation generation after generation, but this gets rather tiring. Instead, we can take a shortcut by noting what happens when we substitute equation (4.1) into equation (4.2):

$$p_2 = p_1(1 - \mu) = P_0(1 - \mu)(1 - \mu)$$
$$= p_0(1 - \mu)^2$$

If we do this for another generation ($t = 3$), we get

$$p_3 = p_0(1 - \mu)^3$$

We can easily see now that these results can be extended to any number of generations, and all we need to know is the initial value of p (p_0), the mutation rate (μ), and the number of generations (t):

$$p_t = p_0(1 - \mu)^t \tag{4.3}$$

Because we have only two alleles, and because $p + q = 1$, the frequency of the a allele in generation t is therefore

$$q_t = 1 - p_t \tag{4.4}$$

For example, if we start with allele frequencies of $p_0 = 1$ and $q_0 = 0$ and a mutation rate of $\mu = 1 \times 10^{-6}$, then, after 10,000 generations, we will have $p_{10,000} = 0.99005$ and $q_{10,000} = 0.00995$.

The higher the mutation rate, the more change will result. Table 4.2 shows frequencies of a mutant allele for different mutation rates at different numbers of generations. For low mutation rates characteristic of most point mutations, there is very little change even after 10,000 generations. For higher mutation rates, there is a greater amount of change, but it is still relatively low for most rates. Only when the mutation rate is very high (10^{-4}) do we see more change, and even then, the accumulation of mutation takes several thousand generations. Eventually, the mutant allele would replace the other allele, but this would take a very long time, even with high mutation rates. These slow rates of change due to mutation are analogous to the slow pace with which a dripping tap will fill a bathtub.

TABLE 4.2   Allele Frequencies under Irreversible Mutation[a]

t	Mutation Rate (μ)				
	10^{-8}	10^{-7}	10^{-6}	10^{-5}	10^{-4}
0	0.00000	0.00000	0.00000	0.00000	0.00000
1,000	0.00001	0.00010	0.00100	0.00995	0.09517
2,000	0.00002	0.00020	0.00200	0.01980	0.18128
3,000	0.00003	0.00030	0.00300	0.02955	0.25919
4,000	0.00004	0.00040	0.00399	0.03921	0.32969
5,000	0.00005	0.00050	0.00499	0.04877	0.39348
6,000	0.00006	0.00060	0.00598	0.05824	0.45120
7,000	0.00007	0.00070	0.00698	0.06761	0.50343
8,000	0.00008	0.00080	0.00797	0.07688	0.55069
9,000	0.00009	0.00090	0.00896	0.08607	0.59345
10,000	0.00010	0.00100	0.00995	0.09516	0.63214

[a]The data presented in this table show the frequency of the a allele (q) under a model where the A allele can mutate into the a allele, but not in reverse. Values were computed for different mutation rates for different numbers of generations (t) using equations (4.3) and (4.4) with starting values of $p_0 = 1$ and $q_0 = 0$.

B.   Reverse Mutation

Before looking at the evolutionary implications of the slow pace of the irreversible mutation model, we should examine what happens when we relax the assumption that mutations are irreversible. Under a model of reverse mutation, we will allow mutation back from the a allele to the A allele as well as the mutation of A to a. As before, we use the symbol μ to represent the probability that an A allele will mutate into an a allele. We will also use the symbol v, the Greek letter nu, to represent the probability that an a allele will mutate back into an A allele. Let us start with p_0 as the initial frequency of the A allele. What will this frequency be after a generation under the model of reversible mutation?

Under the model of reverse mutation there are two ways to obtain an A allele in the next generation. First, a certain proportion of A alleles will *not* mutate. As shown earlier, the probability of an A allele *not* mutating into an a allele is, from equation (4.1), equal to $p_1 = p_0(1 - \mu)$. Because mutation is reversible, we can also get an A allele that has mutated back from an a allele. The probability of this happening is the product of the probability of having an a allele ($= q = 1 - p$) and the probability of an a allele mutating ($= v$), giving the probability $(1 - p)v$. We now have two different ways of obtaining an A allele, so we use the OR rule to figure out the probability of an A allele not mutating *or* an a allele mutating into A. We add these probabilities, giving

$$p_1 = P_0(1 - \mu) + (1 - p_0)v$$

We could then calculate the allele frequency in the next generation (p_2) using the same logic. In general, we can express the allele frequencies in any given generation (t) in terms of the mutation rates and the allele frequencies in the previous generation $(t - 1)$ as

$$p_t = p_{t-1}(1 - \mu) + (1 - p_{t-1})v \qquad (4.5)$$

and the frequency of the a allele as

$$q_t = 1 - p_t \qquad (4.6)$$

The impact of reverse mutation is shown in Figure 4.1, which tracks the frequency of the a allele over 100,000 generations starting with an initial frequency of $q_0 = 0$, a mutation rate of $\mu = 5 \times 10^{-5}$, and a reverse mutation rate of $v = 1 \times 10^{-5}$. This case is contrasted with the case where there is no reversible mutation ($v = 0$). In both cases, the frequency of the mutant allele increases slowly over time (the dripping tap analogy). Under reversible mutation, the increase slows down over time and levels off to a value a bit higher than $q = 0.8$. This happens because an equilibrium is approached between the addition of new a alleles (mutating from A) and the subtraction of preexisting a alleles that mutate back into A. In terms of the dripping tap analogy, consider a leaky faucet that is slowly filling a bathtub, but there is an outlet drain about 80% of the height of the tub. The water continues to enter the bathtub but does not overflow because the incoming water is balanced by the loss of water through the drain.

We can also use this model to illustrate the principle of an equilibrium between counteracting forces, something we will return to a number of times in the text. In this case, we are interested in looking at the equilibrium between forward $(A \rightarrow a)$

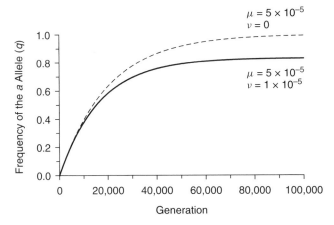

FIGURE 4.1 The frequency of a mutant allele under reversible mutation. The solid line shows changes in the frequency of the a allele (q) starting with an initial value of $q_0 = 0$ with a mutation rate $(A \rightarrow a)$ of $\mu = 5 \times 10^{-5}$, and a reverse mutation rate $(a \rightarrow A)$ of $v = 1 \times 10^{-5}$. Values were computed using equations (4.5) and (4.6). Reverse mutation is contrasted with the case of irreversible mutation (dashed line) with $\mu = 5 \times 10^{-5}$ and $v = 0$. Under reversible mutation, the frequency of q approaches an equilibrium between new mutations and reverse mutations, which is defined as $\mu/(\mu + v) = (5 \times 10^{-5})/(5 \times 10^{-5} + 1 \times 10^{-5}) = 0.8333$ [see equation (4.10) in text].

and backward ($a \rightarrow A$) mutation. We start by looking at the difference in allele frequency between two successive generations. For the A allele, this difference, labeled Δp, is the difference between generation t and generation $t-1$. Using equation (4.5), we obtain

$$\Delta p = p_t - p_{t-1} = [p_{t-1}(1 - \mu) + (1 - p_{t-1})v] - p_{t-1}$$

When we expand all of the terms and cancel out similar terms, we get

$$\Delta p = -p_{t-1}\mu + (1 - p_{t-1})v \qquad (4.7)$$

This is the amount of change in the frequency of the A allele from generation $t-1$ to generation t, reflecting the removal of A alleles that have mutated into a alleles ($-p_{t-1}\mu$) and the addition of new A alleles resulting from backmutation of a alleles [$(1 - p_{t-1})v$]. When these two factors balance each other out, a state of equilibrium has been reached (I am using the general term *equilibrium* here to refer to a state where there is no further change, and not the more specific Hardy–Weinberg equilibrium). We can do the same thing for the frequency of the a allele, defining the change in q as $\Delta q = q_t - q_{t-1}$, which, if we use equation (4.5) and substitute $(1 - q)$ for p, gives, after some algebraic manipulation, the following equation:

$$\Delta q = \mu(1 - q_{t-1}) - q_{t-1}v \qquad (4.8)$$

Here, the addition of new a alleles [$\mu(1 - q_{t-1})$] is balanced by the loss of a alleles through backmutation.

We define equilibrium when there is no change between successive generations, such that $\Delta p = 0$. This equilibrium is a theoretical construct in a mathematical sense, as Δp gets smaller and smaller but technically does not reach 0 (like a curve approaching, but never reaching, an asymptote). As shown in Figure 4.1, for all practical purposes Δp is almost zero by 100,000 generations. We refer to the theoretical equilibrium value, where $\Delta p = 0$, as p_∞, which can be solved by setting equation (4.7) equal to zero, and solving for p, which gives

$$p_\infty = \frac{v}{\mu + v} \qquad (4.9)$$

We can do the same thing for the equilibrium frequency for the a allele by setting equation (4.8) equal to zero and solving for q, giving

$$q_\infty = \frac{\mu}{\mu + v} \qquad (4.10)$$

Finally, we can use the equilibrium values to express the model of reverse mutation in an easier and more precise computational form. Equation (4.5) starts with generation zero and then proceeds to compute each successive generation, one after the other. If, for example, we wanted to compute the allele frequencies after 15,000 generations under the reverse mutation model, we would need to start with generation 0, go to generation 1, and then proceed to generations 2,

3, and so forth, all the way up to 15,000. Although this is easy to do with a computer spreadsheet, there are times when it would be easier to have a formula for plugging in the value of $t = 15,000$. For reverse mutation, the formula is

$$p_t = p_\infty + (1 - \mu - \nu)^t (p_0 - p_\infty)$$

$$q_t = 1 - p_t \tag{4.11}$$

where p_0 is the initial frequency of the A allele in generation 0 and p_∞ is the equilibrium value defined in equation (4.9). For example, using the mutation rates in Figure 4.1 of $\mu = 5 \times 10^{-5}$ and $\nu = 1 \times 10^{-5}$, the equilibrium values would be $p_\infty = 0.16667$ and $q_\infty = 0.83333$. After 15,000 generations starting with allele frequencies of $p = 1.0$ and $q = 0.0$, we get allele frequencies of $p = 0.50547$ and $q = 0.49453$. The importance of this equation is not so much for enabling us to predict allele frequencies for a particular generation but as an example of a more general process of solving iterative equations such as equation (4.5) using what is known as a *recurrence relation*. Details of this approach are outlined in the Appendix 4.1.

C. The Number of New Mutants in a Generation

One thing is obvious from both the irreversible and reversible models of mutation—even when mutation rates are relatively high, the actual amount of change due to mutation alone is very small. Why, then, do we even bother looking at mutation? The answer, of course, goes back to what was stated earlier—mutation is a necessary, but not sufficient, cause of evolution. Evolution is not *just* mutation, but mutation is essential for any evolutionary change because it is the ultimate source of all genetic diversity. Given this distinction, we need to focus not on how mutation changes allele frequencies from one generation to the next (such changes are very, very small), but instead focus on how mutation *introduces* new alleles into a population.

Still, it may appear that mutation could not do much even in terms of introducing new alleles because of the low rate of mutation for any given base pair per individual per generation (Hamilton 2009). To understand the importance of mutation as a source of new variation in a population, we need to shift our attention away from the probability of mutation for a given base pair in a particular individual and instead consider the total number of mutations. The human genome has 3.2 billion base pairs (3.2×10^{-9}), and each individual has two genomes (one from each parent). Given a mutation rate of 2.3×10^{-8} per base pair (Jobling et al. 2004), this means that each person in any generation is expected to have $(2.3 \times 10^{-8}) \times (3.2 \times 10^{-9}) \times 2 = 147.2$ bp mutations. Even if we take a smaller estimate of the mutation rate in humans per base pair of 1.0×10^{-9}, we get 6.4 new mutations for each person in a population (Hamilton 2009). As noted at the beginning of this chapter, statistically *all* of us carry mutant genes and therefore all of us are mutants.

These numbers are even larger when we shift our focus from the individual to the population. Given an estimate of six new mutants per individual, this means that there will be 6000 new mutations each generation in a population of 1000 people. Using the larger estimate of 147 new mutations per person, there will be 147,000 new mutations in a population of 1000 people. At the level of our entire species, this number is even more impressive. As of mid-2010, there were an estimated 6.9 billion people on the planet (Population Reference Bureau 2010). Of these, roughly 3.6 billion were of reproductive age (15–54 years of age for males, 15–44 years of age for females) and form the current reproductive generation (US Census Bureau web page, `http://www.census.gov/ipc/www/idb/worldpop.php`). Using the two different mutation rates given above (6–147 mutations per person), this translates to between about 22 billion and 529 billion new mutations each generation! Regardless of the exact number, it is clear that the *potential* for new genetic diversity is quite high in our species. In fact, it is interesting that the rapid population growth of our species since the beginnings of agriculture some 12,000 years ago has produced a situation where the number of potential new mutations has never been greater (Hawks et al. 2007).

D.   The Fate of Mutant Alleles: An Introduction

Mutation introduces new alleles in a population or species. The other evolutionary forces then affect what happens to these new mutations. The major focus here is on the impact of selection and genetic drift. The interaction between these forces and mutation will be covered in detail in later chapters but for the moment are reviewed briefly here to give the reader a general feel for the broader evolutionary significance of mutation.

The impact of natural selection on mutation depends on how a new mutation affects the likelihood of an individual surviving and reproducing. If a new mutant is harmful, then it will be reduced in frequency. As will be shown in Chapter 6, the rate at which the allele is removed depends on a number of factors, including the different probabilities of survival and reproduction of the different genotypes. As a rule, selection against a mutation will keep the overall frequency of that mutant very low. On the other hand, if a mutant allele is helpful in a given environment, and increases the probability of surviving and reproducing, then the frequency of that mutation will increase over time. A classic example is the case of selection for dark coloration in the peppered moth that was described in Chapter 1. When the dark allele led to a harmful effect (birds were originally more likely to see, and hence eat, dark-colored moths), the frequency remained low. On the other hand, when the environment changed, and dark coloration shifted to being adaptive, then dark-colored moths were selected for, and the frequency of the allele increased.

What happens when selection is not a factor, which will occur when a mutant allele is neutral (i.e., it does not matter what allele you have)? An entire body of literature known as the **neutral theory of evolution** describes how changes in the frequency of a mutant allele are dependent on genetic drift. Various models of neutral evolution will be covered in more detail in Chapter 5. For the moment, we can look at the impact of genetic drift on a mutant allele in general terms. Genetic drift is the random fluctuations of allele frequency with each generation, and has a greater expected effect in small populations (although drift does occur in all populations, we expect a greater amount of drift when populations are small). Drift continues until an allele frequency of 0 or 1 is reached, after which there is only one allele in the population. Because any specific mutation (for a given base pair or locus) is relatively rare, drift will *usually* (not always) result in loss of the mutant allele. For example, imagine that we have a population of 100 people, each with the genotype *AA*, such that we have 200 *A* alleles. If a mutation occurs from *A* to *a*, we now have 199 *A* alleles and 1 *a* allele. We now have 99 people with the *AA* genotype and one person with the *Aa* allele. Suppose that each person in the population has two children. For the *a* allele to make it into the next generation, the person with the *Aa* genotype must pass on the *a* allele to at least one of his or her offspring. By chance, this person might pass on only the *A* allele, in which case the mutant allele is lost. At the level of the population, the same thing occurs. Most of the time, the fate of a new mutant allele is extinction.

However, "most of the time" is not the same as "always." On occasion, the frequency of a mutant allele will drift upward, so that the allele becomes more common in the next generation. Of course, it could then drift down again, perhaps leading to extinction, but some of the time the frequency of the mutant will continue to drift upward, until the new mutation has spread throughout the population, reaching a frequency of 100%! Overall, given enough time, genetic drift will lead to one of two outcomes—the allele becomes extinct (frequency = 0) or fixed (frequency = 1). Although the first outcome is more likely, it is important to note that the second does occur.

III. MUTATIONAL HISTORY AND ANTHROPOLOGICAL QUESTIONS

Although more detail on mutation will be provided in later chapters, it is useful to consider at this point two examples of how the history of mutations in a species or population can inform us about evolutionary questions in anthropology.

A. The Relationship between African Apes and Humans

It has long been recognized that humans and the group of primates commonly known as the "apes" are closely related. This recognition was based initially on anatomical similarities and later confirmed on genetic grounds. Apes and humans are all classified into the superfamily Hominoidea, often referred to simply as *hominoids*. Both apes and humans are classified as hominoids because they share a number of traits not found in other primates, indicating a shared ancestry. Such traits include lack of a tail, a mobile shoulder joint, and presence of scapulae (shoulder blades) on the back as opposed to more on the side of the body, as in monkeys.

A long-standing traditional classification of hominoids recognizes three zoological families, known as the *hylobatids* (also known as the "lesser apes"), the *pongids* (also known as the "great apes"), and the hominids (humans and their ancestors). Figure 4.2 shows this traditional classification. The use of three different families had been very useful for many years because this classification mirrored some of the more obvious anatomical differences. The lesser apes, consisting of gibbons, are small apes that are more monkeylike in some ways. The great

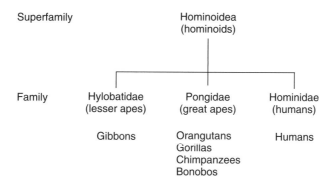

Superfamily   Hominoidea
              (hominoids)

Family   Hylobatidae   Pongidae   Hominidae
            (lesser apes)   (great apes)   (humans)

                Gibbons   Orangutans   Humans
                              Gorillas
                              Chimpanzees
                              Bonobos

FIGURE 4.2 Traditional classification of hominoids. On the basis of certain anatomical features, hominoids have often been classified into three families: hylobatids (lesser apes), pongids (great apes), and hominids (humans). This classification does not agree with evolutionary relationships based on genetic analyses (e.g., Figure 4.3). Consequently, many have argued for a reclassification of the hominoids that better reflects the genetic and evolutionary reality, such as the classification shown in Figure 4.4.

apes are much larger, and consist of the Asian great ape, the orangutan, and the three different types of African great apes—gorillas, chimpanzees, and bonobos (a species that is very closely related to chimpanzees and was once thought to be a subspecies of chimpanzee).

If we assume the traditional classification to be a reflection of evolutionary relationships, we would conclude that the great apes are all more closely related to each other than to humans. We might then infer that there was an evolutionary split at some point in the past between humans and an ancestral great ape, followed later by evolutionary splits between the various species of great apes. Genetic analyses have shown this is not the case. Instead, we now know that African apes and humans are actually more closely related to each other than either is to the orangutan. In other words, there was first a split between the orangutan line and the line leading to a common ancestor of African apes and humans. The closer than expected affinity of African apes and humans was established by immunological comparisons of proteins across different species (Goodman and Cronin 1982). Reactions of proteins found in the serum of blood with antibodies from different species were observed; more distantly related species show greater immunological reaction. As genetic technology increased, other methods of genetic comparison showed the same basic pattern of a closer relationship between African apes and humans than between either and the orangutan. Today, we have the ability to compare the DNA sequences of different primate species, and such comparisons show the same thing. The comparison of DNA sequences further shows that we are more closely related to two of the African apes, the chimpanzee and the bonobo, than to the third African ape, the gorilla.

What does this have to do with mutation? After new species split and form separate evolutionary lines, they no longer share genetic changes. Mutations occur (and are fixed through genetic drift) in one species but not the other. Over time, these mutations accumulate and the species become increasingly different genetically. When we compare immunological reactions, or the amino acid makeup of blood proteins, or direct assessment of DNA sequences, we can generally estimate the total amount of mutational change that has occurred in two species since they shared a common ancestor. The more time that has elapsed since two species have split from a common ancestor, the greater the number of accumulated mutations. Looking at the history of mutations gives us a window on the evolutionary history of different species.

When we describe the number of mutations that separate two species, we are in essence describing the amount of time that has passed since they shared a common ancestor, where time is measured in terms of mutational differences. If we have this information and we have an estimate of how fast such mutations accumulate per generation, we can then translate the number of mutations into a direct estimate of time in generations. This is essentially a distance–rate–time type of problem of the sort we encounter in math classes. You may recall, for example, word problems such as "A car has gone 240 miles at 60 miles per hour. How long has the car been driving?" The answer here is obviously 4 h, based on the simple relationship that distance = rate × time, where in this case we know the distance and the rate and solve for time.

The same principle can be applied to the analysis of mutational differences, a method often known as *molecular dating*, based on the idea of a molecular clock. If we are dealing with DNA sequences or the products of genes (such as proteins) that are neutral, we can use various methods to measure the genetic distance between species, and then estimate the number of generations and the number of years since two species have shared a common ancestor. One way of doing this is to use the fossil record to calibrate the genetic distances. For example, suppose that we have three species (A, B, and C) with the following genetic distances between them: A and B = 12, A and C = 12, and B and C = 4 (the numbers denote units of genetic distance). We see that the distance between species B and C (=4) is less than that between A and B (=12) or between A and C (=12). Thus we can say that B and C are more closely related to each other than either is to species A. Now, suppose that we know from the fossil record that species A split off 30 million years ago. If we assume that the rates of genetic change are the same across all evolutionary lines, then the ratio of time to distance is the same for all species comparisons. Given that species A split off 30 million years ago, and has a genetic distance to other species of 12, we can solve for the time x that species B and C split from each other, resulting in a genetic distance of 4 by assuming that the time to distance ratio is the same, or

$$\frac{30,000,000}{12} = \frac{x}{4}$$

We can now solve for x as $\frac{4(30,000,000)}{12} = 10,000,000 = 10$ million years ago. In reality, the process and methods of estimation are more complex, and involve tests for the accuracy of various assumptions. Nonetheless, this brief hypothetical example gives us not only an example of how mutational history can be used to describe the evolutionary relationship between species but also a rough date of their divergence from common ancestors.

The molecular clock idea was first applied to human evolution by Sarich and Wilson (1967), who used immunological comparison of the albumin protein to infer that humans and chimpanzees diverged from one another approximately 4–5 million years ago (Sarich and Wilson 1967; Sarich 1971). Numerous studies have since been done, using improved calibration from the fossil record and more direct measures of mutational differences from DNA sequences. A typical estimate of the divergence of humans from the common ancestor of chimpanzees and bonobos is now about 6 million years, with an earlier divergence of the gorilla at about 7 million years ago (see Figure 4.3).

A consequence of these genetic studies has been to call the traditional classification of Figure 4.2 into question because some members of the pongid family are actually more similar to humans than to other pongids. Various revisions have been suggested, including the one shown in Figure 4.4, based on the work of Wood and Richmond (2000). Here, the hominid family is broadened to include great apes and humans, and we (humans) are placed in a subfamily (hominines) with our closest relatives, chimpanzees and bonobos, and given our own zoological tribe (a subunit of subfamilies) known as *hominins*. This revised classification has the advantage of showing clearly that African apes and humans are more similar to each other than to the orangutan, and that humans are most closely related to chimpanzees and bonobos. The study of accumulated mutations thus allows us to make precise statements regarding evolutionary history. Keep in mind, however, that although some favor this revised classification because it reflects evolutionary history, others prefer to use the more traditional classification that places humans in

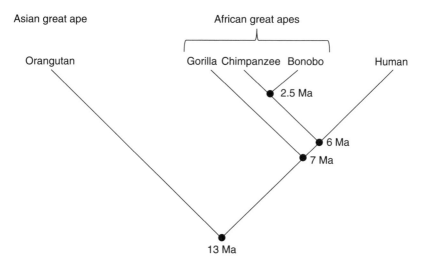

FIGURE 4.3 The evolutionary relationships and estimated divergence dates for great apes and humans based on DNA evidence. All divergence dates are based on nuclear and mitochondrial DNA sequences (Glazko and Nei 2003) except for the split of chimpanzees and bonobos, which is based on mitochondrial DNA (Gagneux et al. 1999). This tree shows that humans and African apes are more closely related to each other than either is to the Asian great ape, the orangutan. In addition, it is clear that humans are somewhat more closely related to the chimpanzee and the bonobo than to the gorilla.

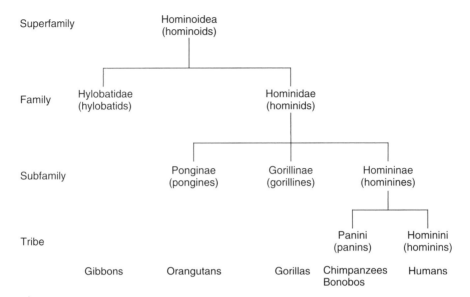

Superfamily — Hominoidea (hominoids)

Family — Hylobatidae (hylobatids) / Hominidae (hominids)

Subfamily — Ponginae (pongines) / Gorillinae (gorillines) / Homininae (hominines)

Tribe — Panini (panins) / Hominini (hominins)

Gibbons   Orangutans   Gorillas   Chimpanzees Bonobos   Humans

FIGURE 4.4 Revised classification of hominoids based on genetic relationships. This classification differs from the traditional classification shown in Figure 4.2 in that it emphasizes the genetic and evolutionary similarities of humans to the African apes, particularly the chimpanzee and bonobo (Wood and Richmond 2000).

a separate family from the great apes (Figure 4.2) because it emphasizes adaptive differences (e.g., Marks 2011). Choice of classification depends on the purpose of the classification.

B. Mutations and Haplogroup Trees

The above example shows how mutational changes in different species can be used to reconstruct the evolutionary history of those species. Here, a different approach is illustrated that provides information on the mutational history *within* a species. The examples used here focus on the human Y chromosome, most of which is inherited solely from father to son without recombination. This means that a man will pass his Y chromosome to his sons intact, who then pass it on to their sons intact, and so on throughout time until a mutation occurs. When a mutation in the Y chromosome occurs, a man with that mutation will then pass it along to all of his male descendants. Over time, such mutations accumulate. When we look at genetic variation among men today, we see the result of this process over many past generations from an initial Y chromosome (the reason why all males can trace their Y chromosome ultimately back to a single male is explained in terms of coalescent theory, which will be described in Chapter 5; for the moment, we are focusing here on what we can learn about the history of past mutations by looking at contemporary genetic diversity).

The process of mutation is shown in Figure 4.5, which traces the genealogy of a single male in generation 0. This male has two sons, each of which has two sons, and so on until there are 16 males in the fourth generation. In order to show

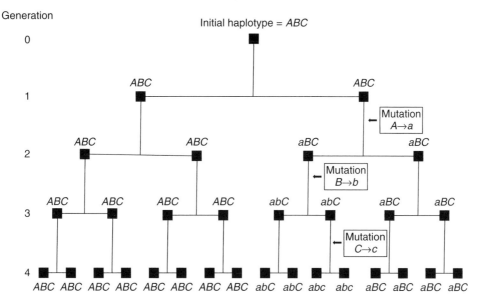

FIGURE 4.5 Tracking mutational changes over time. This hypothetical genealogy of males is used to study changes at three loci in an imaginary Y chromosome, where there are two alleles at each of the three loci: *A* and *a* for the first locus, *B* and *b* for the second locus, and *C* and *c* for the third locus. The genealogy starts with a single male at generation 0 that has the haplotype *ABC*, who has two sons, each of which has two sons, and so on to the fourth generation. Initially, each son inherits the initial haplotype *ABC*, but over time a series of mutational changes occur ($A \rightarrow a, B \rightarrow b, A \rightarrow c$). After each mutational change, all subsequent males inherit that mutation. After four generations, mutation has given rise to three new haplotypes: *aBC*, *abC*, and *abc*. Although the mutation rate here is unreasonably high in order to illustrate several changes in a handful of generations, the principle also applies in the real world where mutations occur less frequently.

the principles of mutational change, we consider a hypothetical situation of three linked loci on the Y chromosome forming a haplotype. The initial male is assigned haplotype *ABC*, which means that he has allele *A* for the first locus, allele *B* for the second locus, and allele *C* for the third locus. He passes his haplotype, *ABC*, to both of his sons. Without mutation, this process would continue forever, with all males in all future generations having the same haplotype. Indeed, this does occur in some of the branches of the genealogy; in fact, 8 in 16 great-great grandchildren still have the *ABC* haplotype.

However, I introduced a mutation in the first locus, from allele *A* to allele *a* in the other line, so that some males in generation 2 have the *aBC* haplotype. This haplotype will then be passed along to all subsequent males until another mutation occurs. Every time a mutation occurs, all subsequent males inherit that mutation. In order to show how this works in Figure 4.5, I set the mutation rate unreasonably high in order to show more mutations in a small amount of time. Consequently, we see three new haplotypes by the fourth generation: *aBC*, *abC*, and *abc*. These haplotypes are related. For example, haplotype *aBC* represents one mutation ($A \rightarrow a$), and haplotype *abC* represents two mutations

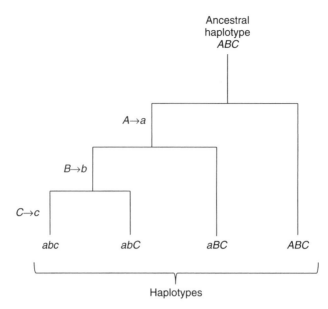

FIGURE 4.6   A haplotype tree summarizing the history of mutations shown in the genealogy in Figure 4.5. Starting with the initial ancestral haplotype of *ABC*, the first change occurs with the mutation *A → a*, giving rise to the haplotype *aBC*. Further mutations occur later in time, generating the haplotypes *abC* and *abc*.

($A \rightarrow a$ and $B \rightarrow b$). Both haplotypes share the mutation of *A* to *a* through descent from a common ancestor, such that haplotype *abC* is descended from haplotype *aBC*. These evolutionary relationships are much clearer when we take this basic information and express it in the form of a haplotype tree, as shown in Figure 4.6. Here, we see the evolutionary connections between the haplotypes, showing that the first mutation was from *A* to *a*, producing haplotype *aBC*. There were then further mutations that affected some of the individuals having this haplotype. The second mutation, from *B* to *b*, produced a new haplotype: *abC*. The third mutation occurred in a male with this haplotype, producing another haplotype: *abc*.

Of course, in the real world there are many more loci and mutations to consider. When we analyze genetic variation in our species, we focus on sets of related haplotypes known as **haplogroups**, which share common mutations. In the example from Figures 4.5 and 4.6, we could, for example, consider all haplotypes with the *b* mutation as part of a single haplogroup, which in this case would include haplotypes *abC* and *abc*. The relationship between haplotype and haplogroup can sometimes be confusing. Elsewhere (Relethford 2003), I use an analogy of automobiles to understand the relationship between haplotype and haplogroup, considering a car's model as analogous to haplotype and a car's make as analogous to haplogroup. Car manufacturers produce a number of different models. Often these models are "related" in the sense that they share certain design features, such as the instrument panel or climate control systems, even though they are different in other ways. In this case, the make of the car, such as Toyota, is analogous to a haplogroup, and the individual models, such as Camry, Corolla, and Matrix, are analogous to haplotypes within the same haplogroup.

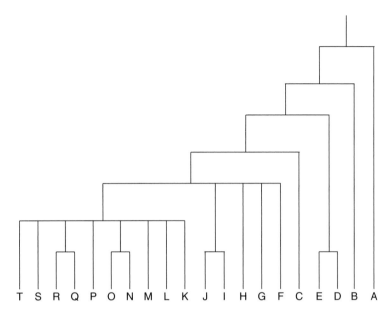

FIGURE 4.7   A haplogroup tree for the human Y chromosome. This tree represents 20 haplogroups, labeled A to T, and is a simplification of the information given by Karafet et al. (2008). Definition of each haplogroup is based on of one or more mutations. For example, haplogroup A is defined as having both the M91 and P97 mutations. This tree is a schematic of major mutational changes in the Y chromosome but is not drawn to scale; the lengths of the branches are not proportional to the ages of the mutational events or the number of mutational events defining each lineage. Of particular interest is the fact that haplogroups A and B are found almost entirely in African populations, suggesting that subsequent evolution of the Y chromosome took place after an expansion out of Africa by modern humans.

In genetic analysis, we often trace the evolutionary relationship between haplogroups. Figure 4.7 presents an example based on global variation in the human Y chromosome. This figure shows the evolutionary relationships of 20 Y-chromosome haplogroups based on a complex type of analysis, known as *parsimony analysis*, that finds the simplest "tree" connecting the haplogroups. The oldest haplogroups are A and B, followed by the common ancestor of haplogroups D and E. Other haplogroups are the result of more recent mutations.

We can learn a lot about past human evolution by examining the geographic location in the world today of the different haplogroups, a method known as **phylogeography**, which looks at the geographic distribution of genetic lineages. Geographic analysis provides clues about the origin and distribution of mutations. A region that has a high frequency of a particular mutation is likely to be the region within which the mutation first appeared. As the mutation spreads through gene flow, it can be found elsewhere, although typically at lower frequency.

In the case of the Y-chromosome tree, we find that haplogroups A and B are almost entirely found only among Africans. Some other haplogroups tend to be restricted to other geographic regions, such as haplogroup D in Asian populations. Some of the other haplogroups are found in several different regions but with high frequencies in one region, such as haplogroup Q, which is found in high frequency in Native American and Northeast Asian populations, but also found at lower frequencies in Europe, the Middle East, and East Asia (Karafet et al. 2008). The fact that the oldest Y-chromosome haplogroups (A and B) are found only in Africa suggests strongly that modern humans arose first in Africa and then spread out to the rest of the world, a finding in agreement with other genetic data and with the fossil record for human evolution (Relethford 2008a). Another interesting finding is the geographic distribution of haplogroup C, which has been found in a number of Asian, Australian, and Pacific populations, but not in any sub-Saharan African populations; this pattern suggests an Asian origin for this haplogroup after the initial dispersion out of Africa. Details of the debate over modern human origins will be discussed later in this book; for the moment, the Y-chromosome data provide us with an example of the insights that can be generated through the analysis of mutations.

IV. SUMMARY

Mutation is a random change in the genetic code. There are various types of mutations ranging from mutations of a single base, to insertions and deletions of larger DNA sequences, to changes in entire chromosomes. Mutations provide the ultimate source of genetic variation, and these new alleles can increase or decrease in frequency as a result of the other evolutionary forces. Mutation rates per locus per individual per generation are very low, and simple modeling shows that mutation alone cannot produce major changes in allele frequencies over even very long periods of time. Although mutation is *necessary* for evolutionary change, it is not a *sufficient* reason, and it requires interaction with other evolutionary forces to produce any significant evolutionary change. Mutation acts to introduce new genetic variation, and the other evolutionary forces act on this new variation to increase or decrease the frequency of mutant alleles. From the perspective of functioning to introduce new variation, the overall impact of mutation is quite noticeable, particularly in large species such as humans today, where billions of new mutations are introduced into our gene pool with each generation. The fate of these mutations is, of course, subject to the impact of other evolutionary forces.

Methods of comparing DNA sequences allow mutation to be used in answering interesting evolutionary questions. One longstanding question in anthropology is the relationship of humans and the great apes. When a species diverges from another, the accumulation of random mutations over time provides us with a genetic yardstick with which to reconstruct evolutionary history. Numerous genetic studies have shown that humans and African great apes (gorilla, chimpanzee, and bonobo) are more closely related to each other than either is to the Asian great ape, the orangutan. This finding has led to the realization that traditional divisions of hominoids into "apes" and "humans" is not entirely accurate, leading to a suggested revision in classification. In addition, genetic comparisons allow us to estimate the date at which different evolutionary lineages split from each other. For example, we estimate that humans split from the African ape lines roughly 6–7 million years ago.

Mutational history can also help us reconstruct the evolutionary history of a species. Genetic comparison allows us to determine the sequence of mutations that have taken place in different haplotypes or haplogroups, leading to an evolutionary "tree" that describes the history of mutation in a species. By examining this tree and the geographic distributions of different haplotypes/haplogroups, we can determine patterns of past migration. An example is the analysis of human Y-chromosome DNA. Globally, there are 20 major haplogroups, of which the two oldest are found exclusively in African populations. The Y-chromosome haplogroup tree fits other genetic and fossil data, suggesting an African origin of modern humans and subsequent dispersion into different geographic regions.

APPENDIX 4.1 USE OF A RECURRENCE RELATION TO SOLVE ITERATIVE EQUATIONS

Many models in population genetics express an allele frequency in terms of the allele frequency in the previous generation, such as equation (4.1) for irreversible mutation and equation (4.11) for reversible mutation. Although sometimes you might want to take derivation of models such as equation (4.11) as a given, it is often useful to understand where they came from. In the case of reverse mutation, equation (4.11) was derived using a recurrence relation, a general model that is applicable to many problems in population genetics. Indeed, some equations in later chapters will be derived (in the appendixes of those chapters) using this relationship. The general principle is outlined here, using reversible mutation as an example.

The following method, modified from Elseth and Baumgardner (1981), applies when we can describe the change in a variable X from one point in time (t) to the next ($t-1$), $\Delta X = x_t - x_{t-1}$, as a first-order linear equation of the form where

$$x_t - x_{t-1} = ax_{t-1} + b$$

By adding x_{t-1} to both sides of the equation, we get

$$x_t = ax_{t-1} + b + x_{t-1}$$

$$x_t = x_{t-1}(a+1) + b \tag{A4.1}$$

We now add the quantity (b/a) to both sides, giving

$$x_t + \frac{b}{a} = x_{t-1}(a+1) + b + \frac{b}{a}$$

We then factor b out from the right-side of the equation and perform several algebraic manipulations, giving

$$x_t + \frac{b}{a} = x_{t-1}(a+1) + b\left(1 + \frac{1}{a}\right)$$

$$x_t + \frac{b}{a} = x_{t-1}(a+1) + \frac{b(a+1)}{a}$$

$$x_t + \frac{b}{a} = (a+1)\left(x_{t-1} + \frac{b}{a}\right)$$

We now subtract (b/a) from both sides, giving

$$x_t = (a+1)\left(x_{t-1} + \frac{b}{a}\right) - \frac{b}{a}$$

When applied to successive time periods, this equation describes an iterative process than can be simplified. Let us start by considering the value of x in generation $t = 1$, starting with an initial value of x in generation $t = 0$ of x_0:

$$x_1 = (a+1)\left(x_0 + \frac{b}{a}\right) - \frac{b}{a} \tag{A4.2}$$

We now consider the next generation ($t = 2$), which gives

$$x_2 = (a+1)\left(x_1 + \frac{b}{a}\right) - \frac{b}{a} \tag{A4.3}$$

If we substitute equation (A4.2) for x_1 in equation (A4.3), we obtain

$$x_2 = (a+1)\left((a+1)\left(x_0 + \frac{b}{a}\right) - \frac{b}{a} + \frac{b}{a}\right) - \frac{b}{a}$$

which simplifies to

$$x_2 = (a+1)^2\left(x_0 + \frac{b}{a}\right) - \frac{b}{a} \tag{A4.4}$$

which is the same form as equation (A4.2) but with the $(a+1)$ term squared. If we extend this to the next generation ($t = 3$) as

$$x_3 = (a+1)\left(x_2 + \frac{b}{a}\right) - \frac{b}{a}$$

and then substitute equation (A4.4) for the value of x_2, we get

$$x_3 = (a+1)\left((a+1)^2\left(x_0 + \frac{b}{a}\right) - \frac{b}{a} + \frac{b}{a}\right) - \frac{b}{a}$$

which simplifies to

$$x_3 = (a+1)^3\left(x_0 + \frac{b}{a}\right) - \frac{b}{a} \tag{A4.5}$$

We see that equations (A4.2), (A4.4), and (A4.5) are all of the same form, allowing us to generalize the equation for any value of t as

$$x_t = (a+1)^t\left(x_0 + \frac{b}{a}\right) - \frac{b}{a} \tag{A4.6}$$

As a sidenote, mathematically oriented readers may be interested to note that the equilibrium value of this iterative equation (where no further change occurs) is equal to $-(b/a)$.

Having outlined the general method, we can now turn to reversible mutation as an example. Recall from equation (4.5) in the main text of this chapter that the frequency of the A allele under reversible mutation is

$$p_t = p_{t-1}(1 - \mu) + (1 - p_{t-1})v$$

which expands to

$$p_t = p_{t-1}(1 - \mu - v) + v$$

where μ is the rate of mutation from A to a and v is the reverse mutation rate from a to A. This model is a first-order linear equation as described above in equation (A4.1) as

$$x_t = x_{t-1}(a + 1) + b$$

where

$$x_t = p_t$$
$$x_{t-1} = p_{t-1}$$
$$a = -(u + v)$$
$$b = v$$

Substituting these values back into equation (A4.6) gives

$$p_t = \frac{v}{u + v} + (1 - u - v)^t \left(p_0 - \frac{v}{u + v} \right)$$

As the quantity $v/(u + v)$ is equal to the equilibrium value p_∞ as shown in equation (4.9), this simplifies to

$$p_t = p_\infty + (1 - u - v)^t (p_0 - p_\infty)$$

which is equation (4.11) in the main (chapter) text.

CHAPTER 5

GENETIC DRIFT

"God does not play dice with the universe." This statement is attributed to the famous theoretical physicist Albert Einstein concerning the claim of quantum mechanics that, at some level, nature is random and best described by probabilities rather than certainty (Natarajan 2008). To many scientists and philosophers, the view that natural processes can have a random component is difficult to accept. In evolutionary biology, however, it has long been recognized that evolution has both deterministic and random components (Mayr 2000).

In the previous chapter we have already seen an element of randomness in the evolutionary process. The occurrence of mutations at a given locus in a given individual in a given generation is a function of probability. Although we cannot predict exactly when and where a mutation occurs, because it is a random process, we can say something about the relative likelihood of a mutation occurring. This is much like a flip of the coin—we do not know whether a given coin will come up heads or tails, but we can state the probability of either event happening ($= \frac{1}{2}$). We can also use probability theory to make some general statements about the likelihood of a set of events occurring. For example, if we flip 10 coins, how many will come up heads? We might be very lucky and get 10 heads. We could also get 9 heads and 1 tail. The other possibilities are 8, 7, 6, 5, 4, 3, 2, 1, or 0 heads. Although we cannot tell beforehand what the specific outcome of any given toss of 10 coins will be, we can derive the probability of the occurrence of any of those outcomes. For example, the probability of getting all heads from 10 coin flips is roughly one in 1000 (0.000977), and the probability of getting 5 heads out of 10 coin flips is 0.246 (see any introductory statistics book for an explanation of how to do this).

Genetic drift is also a random process. Here, allele frequencies can fluctuate from generation to generation because of chance. Under Hardy–Weinberg equilibrium, we expect allele frequencies to remain constant from one generation to the next in the absence of mutation, selection, or gene flow. As noted in Chapter 2, an assumption of Hardy–Weinberg equilibrium is an infinite population size so there is no sampling deviation. In the real world, there are sampling deviations, and allele frequencies can increase or decrease by chance. As with coin flips, we cannot predict beforehand exactly what will happen as a result of genetic drift,

Human Population Genetics, First Edition. John H. Relethford.
© 2012 Wiley-Blackwell. Published 2012 by John Wiley & Sons, Inc.

but the principles of probability allow us to figure out the likelihood of different events occurring.

I. WHAT IS GENETIC DRIFT?

Although it is easy to give the definition of genetic drift as a random fluctuation in allele frequencies over time, it is more difficult to get a feel for what this actually means in an evolutionary context. We will start by considering the effect of sampling in a genetic context and then build to an example of genetic drift.

A. Genetic Sampling

Let us start with a simple example. Assume that there is a locus with two alleles, A and a, and that you have the heterozygous genotype Aa. When you have a child, you will pass along either the A allele or the a allele. Each event has a 50% chance of happening. Therefore, you might expect to pass on the A allele half of the time and the a allele half of the time. Now, suppose that you have four children. How many children do you expect to receive the A allele? Although your expectation would be to have two children with the A allele and two children with the a allele, in reality you could, by chance, have any of the following combinations:

- All four children will have the A allele (none will have the a allele).
- Three children will have the A allele and one child will have the a allele.
- Two children will have the A allele and two children will have the a allele.
- One child will have the A allele and three children will have the a allele.
- Zero children will have the A allele (they will all have the a allele).

Although we do not know beforehand which of these possibilities will occur, we can figure out the *probability* of any of these events happening. To start with, let us consider the case where all four children will receive the A allele. There is only one way for this to happen—the first child receives an A, the second child receives an A, the third child receives an A, and the fourth child receives an A. The probability that each child receives an A is $\frac{1}{2}$, which means that the probability that *all* will receive the A allele is the probability of the first receiving an A AND the second receiving an A AND the third receiving an A AND the fourth receiving an A. Using the AND rule from Chapter 1, we multiply the probabilities to get the probability that all four children receive the A allele as

$$\tfrac{1}{2} \times \tfrac{1}{2} \times \tfrac{1}{2} \times \tfrac{1}{2} = \tfrac{1}{16} = 0.0625$$

We now move to the case where three children receive an A allele and one child receives an a allele. This is a little bit more complicated, because there are four different ways that this could occur: (1) the fourth child could receive the a, (2) the third child could receive the a, (3) the second child could receive the a, or (4) the first child could receive the a. Because the probability of receiving an a allele is the same as receiving an A allele $(= \frac{1}{2})$, the probability of any of these outcomes is $\frac{1}{2} \times \frac{1}{2} \times \frac{1}{2} \times \frac{1}{2} = 0.0625$. These four possibilities and their probabilities are:

$$AAAa \text{ (probability} = 0.0625)$$

$$AAaA \text{ (probability} = 0.0625)$$

$$AaAA \text{ (probability} = 0.0625)$$

$$aAAA \text{ (probability} = 0.0625)$$

To get the overall probability of getting some combination where three children receive the A allele and one child receives the a allele, we need to use the OR rule from Chapter 1 and *add* the probabilities, which gives the total probability of getting 3 A alleles and 1 a allele as

$$0.0625 + 0.0625 + 0.0625 + 0.0625 = 4(0.0625) = 0.250$$

We now move to the case where two children receive an A allele and two children receive an a allele. There are six different ways that can result in two A alleles and two a alleles:

$$AAaa$$

$$AaAa$$

$$AaaA$$

$$aAAa$$

$$aAaA$$

$$aaAA$$

Because each of these possibilities has a probability of $\frac{1}{2} \times \frac{1}{2} \times \frac{1}{2} \times \frac{1}{2} = 0.0625$, the total probability of some any two children having an A allele is $6 \times 0.0625 = 0.375$.

The case where one child receives an A allele and three children receive an a allele is the same probability as the case where three children receives an A allele and one child receives an a allele. There are four ways for this to occur—*Aaaa, aAaa, aaAa,* and *aaaA*—each with a probability of 0.0625, giving a total probability of one A allele and three a alleles as $0.0625 \times 4 = 0.25$. Finally, the case where all four children receive an a allele (*aaaa*) is the same as the probability of all four children receiving an A allele, which is 0.0625. We now summarize the results in the following table:

Number of A Alleles	Probability
4	0.0625
3	0.2500
2	0.3750
1	0.2500
0	0.0625

Note that the total of *all* probabilities adds up to 1.0. Also, note that there is a lot of variation around the *expected* value of two *A* alleles. In fact, the majority of outcomes in this case will *not* have two *A* alleles (probability $= 1 - 0.375 = 0.625$ of having some number other than two *A* alleles). A difference from the expected outcome (two A alleles) is not unexpected if you think about it, because of the nature of probability. Think about a coin flipping analogy; if you flip four coins, you might get 4, 3, 2, 1, or 0 heads.

Deviations from expected values always result from sampling. The expected value for two *A* alleles out of four children is $\frac{1}{2}$, but this expected value will apply all the time only if we are talking about an infinite number of children! (Yes, this seems strange to consider in a biological sense, but it makes perfect sense mathematically.) In any *real* situation, we are sampling from an expected distribution of equal numbers of *A* and *a* alleles. If you have a finite and small number of children, then we will see different numbers of *A* alleles much of the time.

This sampling effect means that you may not pass on your genetic makeup exactly to your offspring. Now, consider this sampling effect happening in an entire population. The result is that the allele frequency among offspring may deviate from the allele frequency among the parents. This is genetic drift, or more precisely *random genetic drift*, because the process is random—sometimes an *A* allele is passed on, and sometimes an *a* allele is passed on.

B. A Simulation of Genetic Drift

In order to see how genetic drift works, we can perform a simple simulation experiment using a set of random numbers (Cavalli-Sforza and Bodmer 1971). For this experiment, we will start with our usual model of a locus with two alleles, *A* and *a*, where the initial allele frequencies are $p = 0.5$ and $q = 0.5$, respectively. Genetic drift will now be simulated for a population of five people. Because each person has two alleles, this means, that we are dealing with 10 alleles. Given a probability of 0.5 for any allele being an *A* allele, how many *A* alleles will we get if we sample 10 alleles? This is analogous to flipping 10 coins to see how many come up heads. Here, however, we are dealing with the probability that an allele present in the parental generation will be passed on to the offspring generation. The expected value (given an initial allele frequency of 0.5) is 5 in 10 *A* alleles. We know, however, that sampling could result in any number from 0 to 10 *A* alleles.

Most often, we use computer programs to simulate the process of genetic drift, which essentially involves a measure of "coin flipping" inside the computer. Here, we will use a different method of simulation by employing a random-number table. Such tables consist of randomly chosen digits (from 0 to 9) and are useful in demonstrating random processes. Here, I used the table of random digits from Rohlf and Sokal (1995), which is a table of 10,000 random digits produced by a computer program. I randomly selected a starting point in the table and wrote out the first 10 digits listed from left to right:

1060633735

We now use these numbers by setting up a rule that matches these digits with the probability of selecting an A or an a allele. Because the initial frequency for A is 0.5, we will use half of the digits (0–4) to represent the A allele and the other half of the digits (5–9) to represent the a allele. An easy way to do this is to note which of the 10 digits represents the A allele by setting those digits (0, 1, 2, 3, or 4) in boldface. This gives us

1060**633**7**3**5

Because 6 of 10 digits are in the range from 0 to 4, this gives us a new allele frequency in the offspring generation of $p = \frac{6}{10} = 0.6$. The allele frequency has changed from 0.5 to 0.6 in a single generation because of genetic drift.

I continued the simulation to the next generation by drawing the next set of 10 random digits from the table:

4681296239

Because the allele frequency in the parental generation is now 0.6, we have to adjust the coding scheme accordingly, letting the digits 0–5 represent the A allele (which is because the range of these 6 of 10 possible digits corresponds to the new probability of 0.6). We now indicate the digits corresponding to the A alleles in boldface, giving

4**6**8**12**9**623**9

There are now five A alleles out of 10, giving a new allele frequency of $p = \frac{5}{10} = 0.5$.

I then extended the analysis to the next generation by picking the next 10-digit string of numbers:

2381536757

Because the parental allele frequency is again $p = 0.5$, we let the digits 0–4 represent the A allele. Setting these in boldface gives

238**1**536757

There are 4 out of 10 alleles that are A alleles, giving an allele frequency of $p = 0.4$. In this case, the allele frequency did not change from one generation to the next.

I continued the simulation using additional random digits over a number of generations. The results are graphed in Figure 5.1. Note that the allele frequency fluctuates over time—sometimes it increases, sometimes it decreases, and sometimes it stays the same. Note that the simulation ends in generation 27 when the allele frequency drifts up from 0.9 to 1.0. Because there are only A alleles in the population, there will be no further change in allele frequency—it will remain at 1.0 unless the a allele is reintroduced through mutation or migration.

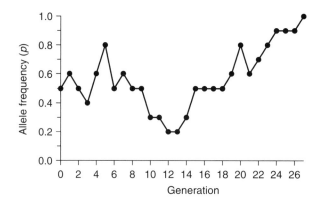

FIGURE 5.1   Computer simulation of genetic drift. This simulation is based on a population size of 5 individuals (10 alleles) and starts with an allele frequency of $p = 0.5$. Random genetic drift is simulated using a table of random numbers as described in the text. Note that the simulation ends at generation 27 when the allele frequency has drifted down to zero. No further change is possible because there are only A alleles in the population. Fixation or extinction of an allele is the eventual outcome of genetic drift.

C.   The Outcome of Genetic Drift

It is important to remember that *each* time you simulate genetic drift you will see some differences. If I were to start the above simulation at a different starting point in the table of random numbers, the run would look different. We can see the random nature of genetic drift by comparing several simulations that all start with the same initial allele frequency and population size. Here, and throughout the remainder of the chapter, the simulations are based on a computer program written to simulate genetic drift. The logic of the program is the same as the simulation above, but the computer is used to generate the random numbers and tally the number of A alleles in each generation (which is *much* faster!).

In order to illustrate the random nature of genetic drift, I performed 100 simulations of drift over 100 generations, each starting with an initial allele frequency of $p = 0.5$ for a population of $N = 50$ individuals (we will continue using the symbol N to indicate population size throughout this text). I selected 3 of the 100 runs were selected to show different outcomes for genetic drift; these are shown in Figures 5.2–5.4. Figure 5.2 shows a case where the allele frequency fluctuates both up and down, and after 100 generations is essentially back to the point where it started. Figure 5.3 also shows fluctuation over time, but eventually drifts up to a frequency of $p = 1.0$ after 78 generations. When the allele frequency reaches this value, it has reached a state of **fixation**; there will be no further change because all of the alleles in the population are A alleles. The only way there could be any more change would be if there were a mutation or if another allele were introduced from another population (gene flow). Figure 5.4 shows a similar outcome, where, after some fluctuation over time, the allele frequency eventually drifts down to a value of $p = 0$ after 64 generations. Here, the A allele has reached a state of **extinction** in that all of the A alleles are gone. No further change will take place unless there is an A allele introduced through mutation or migration.

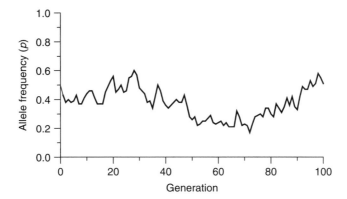

FIGURE 5.2 Computer simulation of genetic drift in a population of $N = 50$ individuals, run 1. The initial allele frequency is $p = 0.5$. The large amount of random fluctuation of allele frequency over time is characteristic of genetic drift in a small population. Compare this graph with two other runs using the same starting parameters as in Figures 5.3 and 5.4.

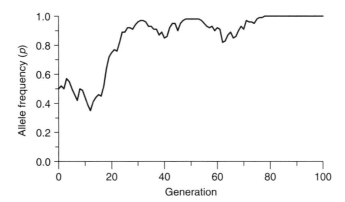

FIGURE 5.3 Computer simulation of genetic drift in a population of $N = 50$ individuals, run 1. The initial allele frequency is $p = 0.5$. The large amount of random fluctuation of allele frequency over time is characteristic of genetic drift in a small population. Note that there is no further change in allele frequency after generation 78, at which point the allele has become fixed in the population. Compare this graph with two other runs using the same starting parameters as in Figures 5.2 and 5.4.

The above examples are meant only to give the reader a taste of some extreme outcomes. In reality, any value between $p = 0$ and $p = 1$ is possible. This brings up an interesting question—although we cannot predict the outcome of any specific case of genetic drift, can we make any predictions about what outcomes are more *likely*? Yes, such predictions can be made using advanced probability theory. Another (and easier to visualize) way of seeing general trends in genetic drift is to use computer simulation of drift over a large number of runs in order to get a visualization of the range of outcomes. Such an example is shown in Figure 5.5, which simulates 100 generations of drift in a population of 50 reproductive adults. Note that

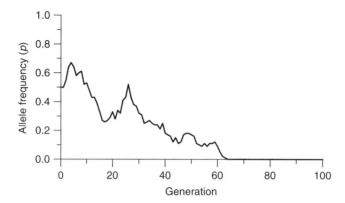

FIGURE 5.4 Computer simulation of genetic drift in a population of $N = 50$ individuals, run 1. The initial allele frequency is $p = 0.5$. The large amount of random fluctuation of allele frequency over time is characteristic of genetic drift in a small population. Note that there is no further change in allele frequency after generation 64, at which point the allele has become extinct in the population. Compare this graph with two other runs using the same starting parameters as in Figures 5.2 and 5.3.

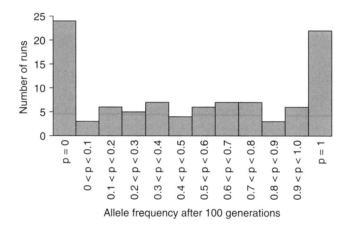

FIGURE 5.5 Computer simulation of genetic drift showing the results of 100 runs of 100 generations of drift in a population of $N = 50$ reproductive adults starting with an initial allele frequency of $p = 0.5$ for each run. Note the tendency for genetic drift to cause the allele frequency to become extinct ($p = 0$) or fixed ($p = 1$) than to have intermediate values. Given enough time, genetic drift will *always* result in extinction or fixation.

the population size here refers only to the number of individuals that are actually reproducing in any given generation (hence the term *reproductive adults*). The distinction between different measures of population size is discussed in more detail later in this chapter; for the moment, it is only necessary to remember that we are counting only those individuals who are reproducing. In addition, we are assuming that the population size stays the same each generation; in other words, 50 adults have 50 offspring that survive to become the adults in the next generation, who then have 50 offspring, and so forth. Again, we will deal with problems with this assumption later, but for now, we will concentrate only on the overall impact of genetic drift.

Because each simulation run is different, we need to conduct a fairly large number of runs to get an idea of average trends. For Figure 5.5, I had the computer program run the analysis from scratch 100 times in order to generate a distribution of outcomes of drift. What should be immediately clear from Figure 5.5 is that many of the runs led to extinction ($p = 0$), similar to what you saw in Figure 5.4, or fixation ($p = 1$), similar to what you saw in Figure 5.3. The rest of the allele frequencies are more or less evenly distributed between the extreme values of $p = 0$ and $p = 1$.

So many of the runs resulted in allele extinction or fixation because genetic drift tends toward extremes over time. In fact, probability theory shows that, given enough time, genetic drift will *always* lead to extinction or fixation. The distribution of allele frequencies after 100 generations shown in Figure 5.5 is well on the way toward the expected end result of *all* runs showing extinction or fixation.

The distribution of allele frequencies changes over time. An example is shown in Figure 5.6, which is based on 1000 simulations of genetic drift in a population of

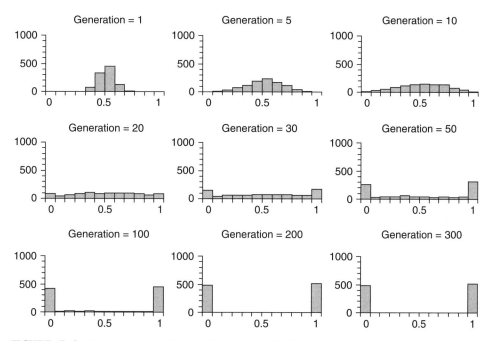

FIGURE 5.6 Computer simulation of genetic drift showing the results of 1000 runs of genetic drift for different numbers of generations in a population of $N = 20$ reproductive adults starting with an initial allele frequency of $p = 0.5$ for each run. The categories for the histograms are those used in Figure 5.5; categories are for 0.1 increment between $p > 0.0$ and $p < 1.0$, with separate categories for $p = 0.0$ (extinction) and $p = 1.0$ (fixation), shown on the left and right sides of the horizontal axis (abscissa). The vertical axis (ordinate) measures the number of runs.

$N = 20$ reproductive adults. Each simulation starts with an initial allele frequency of $p = 0.5$. Figure 5.6 shows the allele frequency distribution at different points in time for the first 300 generations of genetic drift. The first graph (upper left of the figure) shows the allele frequency distribution after a single generation of genetic drift. The allele frequencies all cluster close to the mean (and expected) allele frequency of 0.5, although some runs drifted as low as $p = 0.3$ or as high as $p = 0.775$. By 5 generations, drift has led to a greater spreading of allele frequencies, which increases further by 10 generations. By 20 generations, the distribution is fairly flat, and by 50 generations there is a tendency for the majority of runs to have resulted in either extinction ($p = 0.0$) or fixation ($p = 1.0$). This U-shaped distribution is even more apparent by generation 100 and generation 200, where only a small number of runs resulted in values other than $p = 0.0$ or $p = 1.0$. By 300 generations, *all* of the runs have resulted in either extinction of fixation. According to probability theory, if we ran this simulation an infinite number of times (or at least a very large number of times), half of the runs will result in extinction and half in fixation. The actual numbers in Figure 5.6 are 484 runs that resulted in extinction and 516 runs that resulted in extinction, which is indistinguishable statistically from the expected $50:50$ ratio.

In the above example, we expected that half of the runs would result in extinction and half would result in fixation. The reason for this number ($50:50$) is that the initial allele frequency was 0.5; that is, *half* the alleles were A alleles and half were a alleles. Another way to envision this is to consider a starting point of $p = 0.5$ from which drift will sometimes move to the left of this mean (<0.5) and sometimes will move to the right (>0.5). Each generation, drift will continue to move left or right in a random fashion until the allele becomes extinct ($p = 0.0$) or fixed ($p = 1.0$). Because we started at $p = 0.5$, the distance to randomly drift down to $p = 0.0$ is equal to the distance to randomly drift up to $p = 1.0$. Thus, we expect (subject to sampling error) an *equal* number of cases where extinction and fixation occur.

Given enough time, the ultimate fate of drift is either extinction or fixation, but the relative number of times that each occurs depends on the initial starting value of p. For example, what would we expect if we repeated the same experiment of drift (1000 runs of drift over 300 generations based on a population of $N = 20$ reproductive adults) but started each run with an initial allele frequency of $p = 0.2$? Because drift is a random process, we expect that allele frequencies will drift below $p = 0.2$ and above $p = 0.2$, just as they did when we started with $p = 0.5$. In this case, however, the distance to extinction (moving from $p = 0.2$ to $p = 0.0$) is much less than the distance to fixation (where the allele frequency would have to move from $p = 0.2$ to $p = 1.0$). Thus, even though we would eventually see all possible runs result in either extinction or fixation, we also expect extinction to occur much more often because the initial starting value in this case ($p = 0.2$) is closer to extinction than fixation.

These expectations can be tested using computer simulation. Figure 5.7 shows the results of 1000 runs of genetic drift over 300 generations in a population of $N = 20$ reproductive adults where each run starts at an initial allele frequency of $p = 0.2$. As with our earlier example, Figure 5.7 shows the distribution of allele frequencies after different numbers of generations have passed. Note that over time the allele frequency initially flattens out and then becomes a

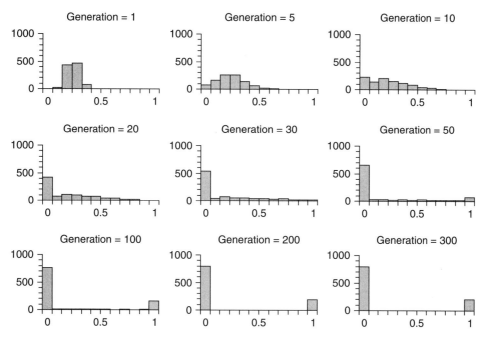

FIGURE 5.7 Computer simulation of genetic drift showing the results of 1000 runs of genetic drift for different numbers of generations in a population of $N = 20$ reproductive adults starting with an initial allele frequency of $p = 0.2$ for each run. The categories for the histograms are those used in Figure 5.5; categories are for 0.1 increment between $p > 0.0$ and $p < 1.0$, with separate categories for $p = 0.0$ (extinction) and $p = 1.0$ (fixation), shown on the left and right sides of the horizontal axis. The vertical axis measures the number of runs.

distribution that has an increasing number of cases ending in allele extinction or fixation. This is what also happened in Figure 5.6, but now there are many more cases of extinction than fixation. After 300 generations have passed, all of the runs have resulted in either extinction or fixation. Unlike the scenario in Figure 5.6, where the number of extinctions and fixations was roughly equal, now there are many more extinctions. Specifically, after 300 generations, we see in Figure 5.7 that 801 runs resulted in extinction and 199 runs resulted in fixation.

The major difference between the simulations shown in Figures 5.6 and 5.7 is that there are fewer cases of fixation in the second set of simulations. The only parameter that was different about these two simulations is the initial allele frequency. In the first set of simulations (Figure 5.6), the initial allele frequency was $p = 0.5$. The relative frequency of fixation in this simulation was 516 runs out of 1000 total runs, giving a rate of $\frac{516}{1000} = 0.516$. In the second set of simulations (Figure 5.7), the initial allele frequency was $p = 0.2$ and the relative frequency of fixation was 199 out of 1000 runs, giving a rate of $\frac{199}{1000} = 0.199$. You may note that in both cases the observed frequency of fixation was almost identical to the initial allele frequency (0.516 vs. 0.5 and 0.199 vs. 0.2). This is not a coincidence, but an expected outcome. In terms of probability theory, the probability of fixation of an allele is equal to the initial frequency of the allele. The small, and statistically insignificant, differences between the simulation experiments and the theoretical expectations are due to sampling error. The expected probability of allele fixation is based on an idealized mathematical model with an infinite number of runs. In the real world, there will be some deviations because we are looking at a smaller number of cases.

As will be described in detail below, the amount of genetic drift is dependent on population size. The above simulation experiments used very small population sizes ($N = 20$) in order to show a lot of drift over the course of 300 generations. If we used a larger population size, it would likely take much longer than 300 generations to reach a state where each run resulted in allele extinction or fixation. In general, the larger the population size, the longer this will take on average (Kimura and Ohta 1969). In a mathematical sense, *any* finite population will *eventually* drift to extinction or fixation. However, in a real-world setting, it could easily take a very large (and unrealistic) number of generations to do so.

II. GENETIC DRIFT AND POPULATION SIZE

In all the above simulations of drift, we needed to consider the number of reproductive adults in the population. The number of breeding individuals in a population determines the extent of likely genetic drift. In short, drift is likely to be greater in a single generation in a small population than in a larger population. The smaller the population, the more likely drift will lead to a larger change in allele frequency. For example, a change in allele frequency in one generation from 0.5 to 0.6 is much more likely for a population of 50 reproductive adults than a population of 500 reproductive adults.

Before considering the impact of population size on drift, it is important to consider exactly what we mean by *population size*. In the context of demography, the size of a population is the total number of people alive at any given point in time. In terms of genetic drift, however, we must only consider those individuals who are actually contributing genetically to the next generation—that is, the number of reproductive adults. We consider the entire population as being made up of three nonoverlapping generations: a prereproductive generation, a reproductive generation, and a postreproductive generation. Individuals that have already reproduced belong to the last generation, and individuals who have not yet reached reproductive age (children) belong to the next generation. If we want to think about the potential for drift in any given generation, we have to focus on the current number of reproductive adults.

Sometimes we contrast these two different views of population size by labeling them **census population size** and **breeding population size**. The former refers to *everyone* in the population, whereas the latter refers to only the reproductive adults. This is an important distinction, because if we want to consider the amount of drift possible in a village of, say, 100 people, we have to remember that the actual breeding population size is much less than 100 after we exclude the younger (prereproductive) and older (postreproductive) individuals. One commonly used rule of thumb is to take breeding population size as one-third of census population size, based on the assumption that a population can be divided into three broad age groups—prereproductive, reproductive, and postreproductive (Cavalli-Sforza and Bodmer 1971).

Although useful for rough comparisons, this measure can be a bit crude as the proportion of individuals of reproductive age in any given population can vary depending on the age structure of the population. For example, among the Dobe !Kung, a hunter–gatherer society of the Kalahari Desert in southern Africa, the number of individuals of reproductive age (15–49 years) in 1968 was 282, out of a total census size of 569, which is almost half (Howell 2000). It is also important to consider cultural influences on age at reproduction. Among the !Kung, women often start reproducing early in their lives—the average age of a mother at the birth of her first child is 19 (Howell 2000). In other cultures, age at first reproduction may be later. In nineteenth–twentieth-century Ireland, for example, it was common for a number of men and women to marry late or to remain single and not reproduce because of a complex interplay of social and economic factors (Kennedy 1973). In addition, as will be described later, many other factors further influence the degree of genetic drift seen in relation to population size. For the moment, we will consider a simple model where we can clearly identify the exact number of reproductive adults in the population (we are also making some other implicit assumptions that will become clear later in this chapter).

A. How Does Population Size Affect Genetic Drift?

The easiest way to see how population size affects genetic drift is to compare a number of runs for different values of population size. Figures 5.8–5.10 do just that. Each of these figures shows the results of five independent runs of genetic drift (to get an idea of the range of likely outcomes) for 100 generations, all starting from an initial allele frequency of $p = 0.5$. The difference between the three figures is the population size used in the simulations. Figure 5.8 uses a population size

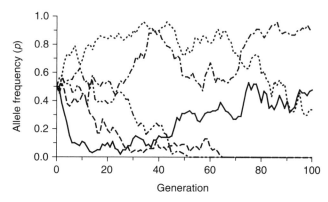

FIGURE 5.8　Five computer simulations of genetic drift for population size of $N = 50$ and an initial allele frequency of $p = 0.5$. The five runs are indicated by different types of lines. Note the wide range of fluctuation in allele frequency. Compare this graph with Figure 5.9 (population size $N = 500$) and Figure 5.10 (population size $N = 5000$).

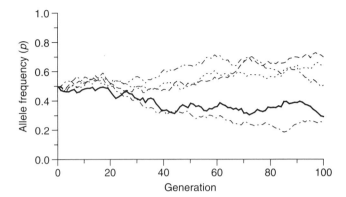

FIGURE 5.9 Five computer simulations of genetic drift for population size $N = 500$ and an initial allele frequency of $p = 0.5$. The five runs are indicated by different types of lines. Note that the range of fluctuation in allele frequency is less than that in Figure 5.8 (for $N = 50$) but still more than Figure 5.10 (for $N = 5000$).

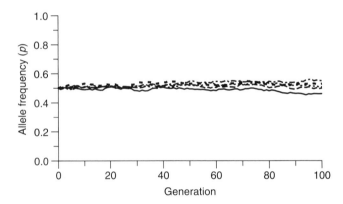

FIGURE 5.10 Five computer simulations of genetic drift for population size of $N = 5000$ and an initial allele frequency of $p = 0.5$. The five runs are indicated by different types of lines. Note that the range of fluctuation in allele frequency is less than in Figure 5.8 (for $N = 50$) and Figure 5.9 (for $N = 500$).

of $N = 50$, Figure 5.9 uses a population size of $N = 500$, and Figure 5.10 uses a population size of $N = 5,000$.

As shown in Figure 5.8, when population size is low ($N = 50$), the allele frequency fluctuates quite a bit from generation to generation. By 100 generations, two of the runs have resulted in allele extinction, one run is close to fixation ($p = 0.9$), and the other two runs have intermediate values ($p = 0.34$ and $p = 0.48$). These results are consistent with the previous simulations (all using small values of N). When we consider Figure 5.9, where the runs are based on a larger population size of $N = 500$, we see that there is still a fair amount of fluctuation in allele frequency over time, but less than was the case for $N = 50$. By 100 generations, the allele frequency ranges from $p = 0.26$ to $p = 0.70$. Moving on to Figure 5.10, where the runs are based on a much larger population size of $N = 5000$, we see that the amount of fluctuation is even more reduced. By 100 generations, the allele frequencies range from $p = 0.46$ to $p = 0.55$.

These simulations show the basic principle of population size and genetic drift—the smaller the population, the greater the amount of drift expected in a generation. The reason for the correlation between population size and drift is statistical sampling, as described in the coin flipping discussions earlier in this chapter. If you flip 10 coins and get 7 or more heads, that is not that unusual. On the other hand, if you flipped 10,000 coins, it would be much less likely to get 7000 or more heads. The more times an event occurs (in this case, the breeding size of a population), the less likely that extreme events occur.

This does not mean that genetic drift occurs only in small populations. As shown in Figure 5.10, there was some drift each generation even when the population size was very large ($N = 5000$). As noted earlier, if we let any drift simulations run long enough, then *eventually* the allele frequencies will become either fixed or extinct (assuming, among other things, that other evolutionary forces, such as mutation, are not operating). This means that, *given enough time*, a population of 5000 will drift to fixation or extinction of an allele. However, the larger the population, the longer, on average, this will take.

Because drift is a random process, an allele could go to fixation or extinction in a very short time or after a very long time. The randomness means that we cannot predict exactly when an allele becomes fixed (or extinct). However, probability theory can give us an estimate of how long the process takes *on average*. Kimura and Ohta (1969) showed that the average number of generations until fixation of an allele (assuming that the allele does not become extinct) $\bar{t}_1$ is

$$\bar{t}_1 = \frac{-4N(1-p)\ln(1-p)}{p} \tag{5.1}$$

where N is the breeding population size, p is the initial allele frequency, and $\ln(1 - p)$ is the natural logarithm of the quantity $(1 - p)$. They also show that the average number of generations until extinction (assuming that the allele does not become fixed) is

$$\bar{t}_0 = \frac{-4N\ln(p)}{1-p} \tag{5.2}$$

In the case of the simulations in Figures 5.8–5.10, the initial allele frequency is $p = 0.5$. In this case, equations (5.1) and (5.2) are the same (because it is the same distance from $p = 0.5$ to fixation or extinction), and reduce to $\bar{t}_1 = \bar{t}_0 = 2.77N$. This means that the average time until fixation or extinction when $N = 50$ is 2.77(50) = 139 generations. For $N = 500$, the average time is 2.77(500) = 1385 generations, and for $N = 5000$, the average time is 13,850 generations. It is clear that fixation or extinction of an allele is more likely in a small number of generations when the population size is small.

If we think about genetic drift in our species, it may be tempting to consider it irrelevant because our species is so numerous (currently about 7 billion people). This would be an incorrect conclusion because our species is not one big population within which everyone has an equal chance of mating, but instead is made up of thousands of smaller breeding populations. Our species is subdivided by geography, ethnicity, language, and many other demographic and cultural factors. In addition, the large species we have today is a recent development. Our species size has increased rapidly in just a few centuries—there were less than 2 billion people only 100 years ago (Weeks 2005), and likely no more than about 6 *million* only 12,000 years ago (Weiss 1984). Given that our ancestors were all hunters and gatherers before 12,000 years ago means that local breeding populations were likely small, perhaps tribes of 200 reproductive adults consisting of a number of smaller bands (Eller et al. 2004). It seems quite clear that there was ample opportunity for genetic drift throughout human evolution, which means that much of our current patterns of genetic diversity likely reflect past genetic drift to a large extent.

One of the most comprehensive studies of genetic drift in human populations has been conducted by Cavalli-Sforza et al. (2004) on populations in the Parma Valley in Italy. The parishes in the valley have a great deal of variation in population size depending on their location, with populations in the mountain regions having the lowest population density, populations in the hills having higher density, and populations in the plains having the highest density. Data were collected on various red blood types and used to measure the amount of genetic variation between groups. They then compared the observed patterns of genetic variation with those expected under a model of genetic drift (using a type of computer simulation more complex than the ones used in this chapter), and found that the majority of genetic variation could be explained by genetic drift, and consequently differences in population size.

B.   Effective Population Size

Estimating the effects of genetic drift gets more complicated once we remember that the simulations used here rely on a simple definition of population size that corresponds to the number of reproductive adults in the population. The model of genetic drift underlying these simulations made a number of other implicit assumptions:

1. We assume that the number of males and females is the same, so that when we say that a population size is $N = 20$ adults, this actually means that there are 10 males and 10 females.

2. We also assumed that population size remained the same from one generation to the next, so that 20 adults gave birth to 20 children who made up the next generation. Furthermore, we also assume that there variation in the number of offspring for each couple is random (that is, described by a statistical distribution known as the Poisson distribution).

It is common when using mathematical models and simulations to make a number of simplifying assumptions. The trick is to determine whether these assumptions are critical and what happens to our basic model if, in the real world, these assumptions are violated. What happens, for example, to genetic drift when population size changes over time? What happens if there are more females than males? How can we deal with violation of our assumptions and investigate the genetic impact of these violations?

Population geneticists have come up with a very ingenious way of dealing with deviations from the simplifying assumptions of our simple model of genetic drift. Instead of using breeding population size (N) directly, we can modify it to incorporate various complexities, producing what we call an **effective population size** (denoted N_e). Effective population size is the breeding population size in an idealized population where a number of conditions, such as equal sex ratio, constancy in population size, and random variation in fertility, apply. An effective population size is an adjusted value of population size that takes these assumptions, such as unequal numbers of males and females or changes in population size over time, into account. Several examples are shown below. Other factors affecting effective population size, as well as the derivations of the effective population size formulas, can be found in a number of advanced texts, such as Wright (1969) and Crow and Kimura (1970).

Changes in Population Size

A good example of the contrast between breeding population size and effective population size is the case of changing population size. So far, we have assumed that population size remains constant from one generation to the next. In the real world, population size often changes, sometimes with minor fluctuations and sometimes with major increases or decreases in population size. As an example, consider a hypothetical population that starts with 200 adults and then declines rapidly for a few generations and then recovers. Given values for six generation of $N = 200, 100, 50, 50, 100, 200$, how much genetic drift should we expect to have resulted over these six generations? Can we come up with a single value of N that takes this fluctuation into account?

After thinking about this, you might conclude that a *mean* (average) value of population size would be a good value to use. This answer is correct, but it turns out that the usual type of average we are most familiar with (the arithmetic mean) is not appropriate here. Instead, the effective population size used when there are changes in population size is the harmonic mean, which is an average of the reciprocals. The effective population size is computed as

$$N_e = \frac{t}{\Sigma\left(\dfrac{1}{N}\right)} \tag{5.3}$$

where the values of $1/N$ are summed (Σ) over all generations and t is the number of generations. For the hypothetical data above, equation (5.3) gives the effective population size as

$$N_e = \frac{6}{\frac{1}{200} + \frac{1}{100} + \frac{1}{50} + \frac{1}{50} + \frac{1}{100} + \frac{1}{200}} = \frac{6}{0.07} = 85.7$$

which can be rounded off to $N_e = 86$. This means that the amount of genetic drift expected from a population that has a constant effective population size of 86 is the same as a population that changes in size from 200 to 100 to 50 to 50 to 100 to 200. Thus, the changes in population size are accommodated with computation of the effective population size that is used in a model assuming constant population size.

The above case is an example of a population that has experienced a **bottleneck**, a dramatic reduction in population size. In this case, the population recovered from the bottleneck, but the effective population size remained lower than the end value of $N = 200$. The effective population size after a bottleneck will always be closer to the minimum value (in this case, $N = 50$) than to the arithmetic average ($N = 117$). Because of the increased effect of genetic drift during a bottleneck, genetic diversity in a population can be reduced substantially. This effect is often seen most dramatically when a small number of individuals find a new population; this loss of genetic diversity is known as the **founder effect**.

A good example of bottlenecks in a human population comes from Roberts' (1968) study of the island of Tristan da Cunha located in the southern Atlantic Ocean settled by the United Kingdom in 1816. From small initial numbers, the population began growing during the nineteenth century, but experienced two severe bottlenecks. The first, occurring in the mid-1850s, resulted from a large number of people leaving the island, and the population declined from 103 to 33. Afterward, the population began to grow again, but a second bottleneck occurred in 1885 when 15 men died in a boating accident. This second bottleneck left only four adult men on the island. Using genealogical records, Roberts was able to trace genetic contributions of initial founders to the "present-day" (then 1961) gene pool of the island, and found that the two bottlenecks resulted in a loss of genetic variability.

Variation in Fertility

One simplifying assumptions that we make with genetic drift is that variation in the number of offspring is random. When this is not the case, and variation is more than expected at random (predicted using what is known as a Poisson distribution), then the effective population size will be less than the breeding population size. There are several measures of effective population size when there is variation in the number of offspring, including complex models that allow for different levels of variation for male and female parents (for situations where males have children with different women). Some measures also allow for populations to change in size over time. Here, we will look briefly at one of the simpler measures just to get a feel for the effect that variation in offspring number can have.

For cases with separate sexes (which includes humans) and a constant population size (where the mean number of offspring per couple is 2), the effective population size is

$$N_e = \frac{4N - 4}{2 + V} \tag{5.4}$$

where N is the breeding population size and V is the variance in offspring number (Wright 1969). Variance is a statistical measure that estimates the average squared deviation from a mean; a more detailed explanation of how variance is computed is given in any introductory statistics text. For the moment, we can interpret V as a measure of how much variation exists in the number of offspring that each couple has; if V is low, then most couples have a similar number of offspring, and if it is high, then there is a greater range. When we use breeding population size to estimate effects of drift, we are assuming that the variance is equal to the average number of births. When V is larger, effective size is reduced.

For example, consider a population of $N = 100$ adults that is constant in size with a variance of $V = 3$. Using equation (5.4), we see that the effective size is $N_e = [4(100) - 4]/(2 + 3) = 79$. If $V = 5$, the effective population size is $N_e = 57$, and if $V = 10$, then the effective size is $N_e = 33$. It is clear that when there is a high level of variation in the number of offspring, then effective population size can be much less than the breeding population size. In some human populations, such as the Irish Travellers (an itinerant population in Ireland), variance in the number of offspring can be quite high (Crawford and Gmelch 1974).

Sex Ratio

Our final example of factors affecting effective population size considers the number of adult males and females in the breeding population. We have been assuming that there are equal numbers of adult males and adult females. If there are more of one sex than another, the effective population size will be less than the breeding population size. Here, effective population size is

$$N_e = \frac{4N_M N_F}{N_M + N_F} \tag{5.5}$$

where N_M is the number of adult males and N_F is the number of adult females. In many cases, the effect of unequal sex ratio is minor in human populations. For example, Salzano et al. (1967) reported 522 males and 547 females between ages 15 and 30 in a survey of the Caingang Indians of Brazil. Taking this age group as an approximation of the breeding population gives a breeding population size of $522 + 547 = 1069$. Using equation (5.5) to adjust for the unequal sex ratio gives $N_e = [4(522)(547)]/(522 + 547) = 1068$, which is essentially the same value. As noted by Gillespie (2004), effective population size is not affected by unequal sex ratio unless the sex ratio is extreme, which is not the case for humans.

Other primates, however, have different social structures, where the number of males and females can be quite different. One example are baboons, Old World monkeys whose social groups typically have many more adult females than adult males. I was once involved in simulations of the genetic structure of a baboon breeding colony in San Antonio, Texas (Relethford 1981). At that time, there were approximately 30 adult male baboons and 300 adult females in the breeding colony, giving a breeding population size of 330. The effective size, as computed using equation (5.5), is $N_e = 109$, which is about a third of the breeding effective size. The unequal sex ratio makes the effective size much smaller.

III. EFFECTS ON GENETIC VARIATION

Ultimately, genetic drift leads to the extinction or fixation of an allele (assuming no mutations or alleles introduced from other populations). As such, the amount of genetic variation within a population will decrease over time as a result of genetic drift. At this point, it is useful to introduce a measure of genetic variation that will be used several times throughout the book.

A. Measuring Genetic Variation

If there is only one allele at a locus, there is, by definition, no variation. If there are two alleles in the population, there is more variation, and if there are three alleles, there is even more variation. Variation also depends on the frequency of the alleles; two alleles each having a frequency of 50% shows more variation than if one allele had a frequency of 99% and the other had a frequency of 1%. The reason for this is more apparent if we consider one widely used measure of genetic variation—**heterozygosity**, which is simply the proportion of heterozygotes in a population. For a locus with two alleles and frequencies of p and q, respectively, heterozygosity is simply the proportion of heterozygotes expected under Hardy–Weinberg equilibrium:

$$H = 2pq \qquad (5.6)$$

The maximum heterozygosity when there are only two alleles is $H = 0.5$, which occurs when $p = q = 0.5$. If there are more than two alleles, heterozygosity is derived as

$$H = 1 - \sum p_i^2$$

where the summation is over all alleles (each allele is denoted by the subscript i, such that the first allele corresponds to $i = 1$, the second to $i = 2$, etc.). For example, if there are three alleles in a population with frequencies of 0.6, 0.3, and 0.1, the heterozygosity for this locus is

$$H = 1 - (0.6^2 + 0.3^2 + 0.1^2) = 1 - 0.46 = 0.54$$

The maximum heterozygosity increases as the number of alleles increases, and approaches a maximum value of $H = 1$ (the limit if there were an infinite number of alleles).

In a real-world analysis, heterozygosity would be computed using many different loci, which involves taking the above measure and averaging it over each locus. The important thing to remember is that heterozygosity is a measure of variation, and the larger the number, the greater the genetic variation in the population.

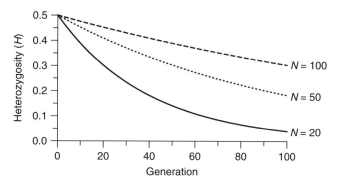

FIGURE 5.11 Decay of homozygosity over time due to genetic drift. Heterozygosity in each generation was derived using equation (5.7) by applying an initial heterozygosity in generation $t = 0$ of $H_0 = 0.5$ for different values of N. The decay over time is faster when population size is smaller.

B. The Decay of Genetic Variation over Time

As noted above, genetic drift leads to extinction of fixation of alleles, which reduces the level of genetic variation in the population. When there is only one allele ($p = 1$ or $q = 1$), then heterozygosity will equal 0. The expected loss of heterozygosity in a population due to t generations of genetic drift is

$$H_t = H_0 \left(1 - \frac{1}{2N}\right)^t \tag{5.7}$$

where H_0 is the initial heterozygosity in generation $t = 0$. Derivation of this equation is presented in Appendix 5.1.

This equation tells us two things: (1) as t increases, heterozygosity gets smaller, approaching a limit of $H = 0$; and (2) the smaller the population size, the greater the reduction in heterozygosity from one generation to the next. The effect of population size on the rate of decrease in heterozygosity due to drift can be seen in Figure 5.11, which shows this decline for three different values of population size.

The loss of heterozygosity due to genetic drift is similar to the loss of heterozygosity (and the increase in homozygosity) seen in inbreeding. Genetic drift can be seen as a random form of inbreeding. Even if mating is at random in a population, if a population is small, then there is a shortage of available unrelated mates, so even a mate chosen at random is likely to be related. More details on this correspondence are given in Appendix 5.1. Although both inbreeding and genetic drift increase homozygosity, there is a difference in that genetic drift also leads to a change in allele frequency, whereas inbreeding does not.

IV. MUTATION AND GENETIC DRIFT

We have so far discussed two evolutionary forces, mutation and genetic drift. Although it has been convenient to discuss them separately, in the real world these two evolutionary forces occur at the same time. Mutation and drift have opposite effects on variation within a population. Drift removes variation in a population, while mutation increases variation by introducing new alleles into the population. Two questions emerge:

1. What happens to a mutant allele under drift?
2. How do mutation and drift interact to affect genetic variation in a population?

A. The Fate of a Mutant Allele

The simulations used in the first part of this chapter all began with an initial allele frequency of $p = 0.5$, but the general method can be used for any initial allele frequency. One example that is particularly revealing is to use simulation to explore the fate of a mutant allele. Consider a population of N adults, each having two copies of the A allele. Now, imagine that one of these alleles mutates into a new allele, a. Since there are N adults, there are $2N$ total alleles, of which one is a new mutant. This makes the initial allele frequency of a new mutant equal to $1/2N$. What do you suppose will happen to this mutant allele under genetic drift?

Because $1/2N$ is a low value and pretty close to zero in most cases, we would expect that most of the time genetic drift would act to remove the mutant allele from the population. This is because there is a small difference between 0 (extinction) and the initial frequency of the mutant allele ($1/2N$). However, we also know that sometimes drift will cause the mutant allele to increase in the next generation. Here, too, the increase is often likely to be small and the new allele could be lost quickly after another generation or so. In *some* cases, however, it is possible that the mutant allele would increase enough by drift to eventually reach fixation ($p = 1.0$). Given what we have already seen, we can predict the likelihood of this happening. Remember that the probability that an allele will eventually reach fixation is equal to the initial allele frequency, and the probability of extinction is 1 minus this frequency. Thus, if the frequency of a mutant allele in a population is 0.01, then there is a 1% chance that this mutant will eventually become fixed, and a 99% chance that the mutant will eventually become extinct. The fate of most mutant alleles is extinction, *but not all*.

We can use simulation to test these predictions. Here, I used a population of $N = 20$ adults where one allele [out of $2N = 2(20) = 40$ alleles] is a new mutant. The initial allele frequency was set to $\frac{1}{40} = 0.025$. Given enough time to reach fixation or extinction, we expect that the proportion of runs where the mutant allele frequency reaches fixation will be 0.025 and the proportion of runs where the mutant allele becomes extinct will be $1 - 0.025 = 0.975$. I ran 1000 simulations for 300 generations of drift using a population size of $N = 20$ and an initial allele frequency of 0.025. As expected, most of the runs resulted in extinction of the mutant allele (971 of 1000 runs = 97.1% of the time, statistically the same as the expected value of 97.5% of the time). Most of the time extinction occurred in a small number of generations. However, in 29 of 1000 runs, the mutant allele actually drifted to reach fixation, a proportion not statistically different from the expected 25 in 1000 runs. Figure 5.12 shows 3 of the 29 runs that resulted in fixation of the

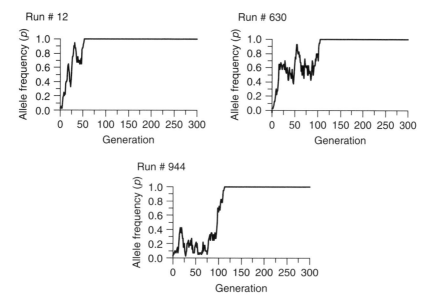

FIGURE 5.12 Fixation of a mutant allele due to genetic drift. As described in the text, computer simulation was used to examine the probability of extinction or fixation of a single mutant allele. In total, 1000 simulations of 300 generations of genetic drift were performed using a population size of $N = 20$. Given a single mutant allele to start with, the initial allele frequency was set as $p = \frac{1}{40} = 0.025$ (because there are $2N = 40$ alleles). In 971 runs, drift led to the extinction of the mutant allele. In 29 runs, drift led to the fixation of the mutant allele. This figure shows three of the runs in which fixation occurred.

mutant allele just to get a feel for how drift will *sometimes* result in a mutant allele spreading through an entire population. Just by chance.

B.   Equilibrium between Mutation and Genetic Drift

Mutation and genetic drift interact to affect the level of genetic variation in a population. As noted earlier, genetic drift acts to remove variation from a population. Mutation, on the other hand, acts to increase variation in a population by adding new alleles. The interaction between mutation and drift can be considered as a balance between alleles being added to the population (mutation) and alleles being lost from the population (drift). This balance can be visualized using a physical analogy such as a paper cup (e.g., Relethford 2001). As shown in Figure 5.13, take a paper cup and punch a small hole in the bottom (about $\frac{1}{4}$ in. works well). Hold the cup under a water faucet that has a small dribble of water. The water enters the cup and then exits through the hole in the bottom. If you play with the rate of water entering the cup, you will find a rate where the water begins to rise in the cup to a certain level where the amount of water entering the cup is balanced by the amount of water leaving the cup.

Mutation and genetic drift in a population can be regarded in the same way; there is a level of genetic diversity in a population where the introduction of new

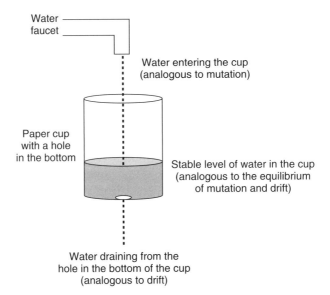

FIGURE 5.13 Graphic analogy of the balance between mutation and drift. Water dribbles slowly from a faucet into a paper cup that has a small hole in the bottom, through which the water exits the cup. If you adjust the flow of water to just the right rate, you will get a balance between the water entering the cup and that exiting the cup. This example is a physical analogy of the balance between mutation and genetic drift.

genetic diversity through mutation is balanced by the loss of diversity due to drift. This balance, or equilibrium, can be seen in mathematical terms. Here, we use what is known as the **infinite alleles model**, a simplification that assumes that each new mutation results in a brand new allele that does not already exist in the population. For an effective population size of N and a mutation rate of μ, the expected heterozygosity when an equilibrium is reached between mutation and drift is approximately

$$H = \frac{4N\mu}{1 + 4N\mu} \tag{5.8}$$

Derivation of this equation is presented in Appendix 5.2. The quantity $4N\mu$ determines the relative influence of mutation and drift. When $4N\mu$ is very small (close to 0), then the expected heterozygosity at equilibrium will be close to 0, typical of a very small population experiencing heavy genetic drift. When $4N\mu$ is very large (>5 or so), the effect of genetic drift is minimal and the expected heterozygosity is high (>0.83). Under the infinite alleles model, the maximum heterozygosity is $H = 1.0$. Intermediate values at or near $4N\mu = 1$ show an expected balance between mutation and drift (Hamilton 2009). The balance between mutation and drift at equilibrium forms an important part of the neutral theory of evolution that was introduced in the last chapter. The neutral theory sees much of allele frequency change as genetic drift operating on neutral mutations (those that do not affect survival or reproduction).

The balance between mutation and drift has also been extended to molecular data where we look at differences between the DNA sequences of individuals using the **infinite sites model**. Here, we consider the probability of a mutation at any particular nucleotide site to be low enough that each new mutation along a DNA sequence will occur at a different location (i.e., the same nucleotide site is unlikely to mutate more than once). If we look at sufficiently long sequences of DNA, the probability of multiple mutations is low enough that we can consider this very long sequence to be practically infinite. Variation in a sample of DNA sequences occurs when there is a different nucleotide at the same position in different DNA sequences. The average proportion of differences between all sequences in a sample is known as **nucleotide diversity**, typically denoted by the Greek letter π(pi). Appendix 5.3 contains more details on how this measure is computed. The important thing to keep in mind for the moment is that the expected level of nucleotide diversity at equilibrium between mutation and drift is the same as equation (5.8), where the mutation rate is defined as the total probability of mutation across the entire nucleotide sequence (Hartl and Clark 2007).

V. COALESCENT THEORY

Thus far, we have been considering genetic drift as a process that goes from the past to the present, where we are concerned with the probability that a given allele will be represented in future generations. On an individual level, consider what could happen to an allele that you have for a given locus. Imagine that you have two children. The possible outcomes are that you would pass on two, one, or no copies of this allele (just as when flipping a coin twice in a row, you can get two, one, or no heads). The simulations used so far in this chapter basically use this idea extended to an entire population. Although standard population-genetic models of drift use this "forward" approach from past to present, we can also look at the process of drift in the opposite direction, from present back into the past. Instead of considering the probability that a given allele will be represented in a future generation, we instead consider the probability that any two alleles share the same ancestor in a previous generation. This "backward" approach, from the present back into the past, forms the basis of a powerful set of models known as **coalescent theory**.

In order to show how coalescent theory works, we start with a graphic representation of the standard forwards interpretation of genetic drift. Figure 5.14 shows genetic drift starting with five different alleles four generations in the past (the five circles shown at the top of the figure labeled 4a, 4b, 4c, 4d, and 4e). Imagine that each generation a given allele can give rise to two, one, or no descendants, and the actual outcome is a matter of probability (determined, e.g., by flipping coins). In Figure 5.12, the first three alleles (4a, 4b, and 4c) all give rise to a single descendant allele each, while the fourth allele (4d) gives rise to two descendants (3d and 3e). Finally, allele 4e (farthest to the right on the top row) leaves no descendants—it has become extinct. We now follow the process over the course of four more generations until we get to the present day (generation 0). Looking over this simple simulation, we see that there are five alleles in each generation,

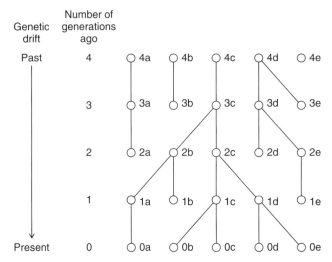

FIGURE 5.14 Genealogy showing genetic drift and the extinction of alleles over time. We start with five alleles four generations in the past. Circles represent alleles and lines indicate descent. Each allele has been labeled with a number and letter (e.g., 0c, 3b) for reference, where the number refers to the number of generations in the past. With each generation, an allele can leave two, one, or no copies in the next generation. For example, from four generations ago to three generations ago, alleles 4a, 4b, and 4c each left one descendant allele in the next generation, whereas allele 4d left two descendant alleles and allele 4e left none. When no alleles remain, the allele becomes extinct (alleles 4e, 3b, 3e, 2a, 2d, 1b, 1e). The five alleles in the present day are all descended from a single allele (3c) three generations earlier.

but along the way, some alleles are represented in subsequent generations, and some alleles become extinct along the way.

If you look more closely at Figure 5.14, you will see that all five alleles in the present day are descended from a single allele (3c) three generations earlier (you can trace the lines of descent back from the present day to see this). In turn, this common ancestral allele is descended from allele 4c in the previous generation. Coalescent theory looks backward in time from the present day, and is concerned only with those lines that trace back to a common ancestor; we do not consider alleles that become extinct. In order to demonstrate this principle, Figure 5.15 duplicates Figure 5.14 but with an important exception—the alleles that lie in the lines of descent back from the present day to common ancestors are represented by filled circles, and lineages of extinct alleles are represented by open circles. If we start at the present day and work backward, we can see two cases where two alleles coalesce to a common ancestor in the previous generation—alleles 0b and 0c coalesce to allele 1c, and alleles 0d and 0e coalesce to allele 1d. For Figure 5.15, each generation back in time from the present shows additional coalescent events. By three generations in the past, the lineages leading backwards from the five alleles in the present day have all coalesced to a single allele (3c). In coalescent theory, we call this single allele the **most recent common ancestor**, which is often abbreviated as MRCA. We refer to this ancestor as the "most recent" because this

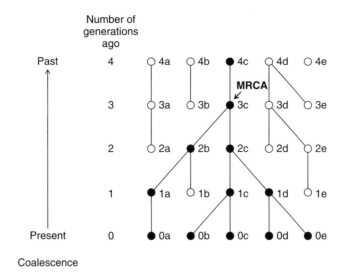

Number of
generations
ago

Past   4

3

2

1

Present   0

Coalescence

FIGURE 5.15 The process of coalescence. The genealogy used in Figure 5.14 is used, but with filled circles representing alleles that trace their ancestry back four generations earlier, and open circles representing lineages that became extinct. This version of Figure 5.14 shows more clearly how all five alleles in the present day trace their ancestry back to a single allele three generations in the past. Going back in time from the present day, we can see how lines of ancestry *coalesce* into single alleles; for example, the second and third alleles in generation 0 (alleles 0b and 0c) coalesce into a single allele in the previous generation (1c). The process of coalescence continues until the lines of ancestry for all five alleles have coalesced into a single allele, marked on the diagram as the most recent common ancestor (MRCA), which is allele 3c. Coalescent theory is used to make inferences about the MRCA, but cannot tell us about the ancestor of the MRCA (4c, the ancestral allele in the top row).

is the first allele we find going back in time that is ancestral to all later alleles. Note that the most recent common ancestor (3c) also has an ancestor (allele 4c). However, in coalescent theory we can make evolutionary inferences back only as far as the most recent common ancestor.

A. Average Time to Coalescence

Coalescent theory starts by looking at the probability that two alleles came from a single allele in a previous generation. Coalescence theory extends these concepts to more than two distinct alleles. For a population of k distinct alleles, there is a point in the past where these k alleles coalesce into $k - 1$ alleles, and a point earlier in time when these alleles in turn coalesce into $k - 2$ alleles, and so on. For example, if we had a population in the present day with six distinct alleles, there must be a point in the past when these six alleles coalesce into five alleles. In turn, these alleles then coalesce into four, then three, then two, and then one allele (the most recent common ancestor).

Coalescent theory is concerned with the probability of common ancestry. If we examine two alleles in any given generation, what is the probability that they both came from the same allele in the previous generation? If we assume a population of constant size $2N$, the probability that an allele in a given generation will have a specific allele in the previous generation is $1/2N$. The probability that another allele will also have this same ancestral allele is $(1/2N)(1/2N)$, based on the AND rule. However, this probability applies only to common ancestry of a *specific* allele in the previous population, and there are $2N$ alleles in the previous generation. In order to figure out the probability of *any* allele in the previous generation being a common ancestor, we have to use the OR rule and consider the total probability over all alleles, which gives

$$(1/2N)(1/2N)2N = 1/2N$$

It also follows that the probability of two alleles *not* having the same ancestral allele in the previous generation is

$$1 - (1/2N)$$

This can be extended into the past. The probability that two alleles in the present had a common ancestor $t+1$ generations ago is

$$\frac{1}{2N}[1 - (1/2N)]^t$$

(Hartl and Clark 2007:130).

An important quantity in coalescent theory is the average number of generations it takes until coalescence of two alleles. Probability theory allows an estimate of the average time until coalescence [the derivations and other mathematical background are beyond the scope of this text—see advanced texts such as Hartl and Clark (2007) and Hamilton (2009)]. For a diploid population of size N and k distinct alleles, the mean number of generations until coalescence is

$$\frac{4N}{k(k-1)} \tag{5.9}$$

For example, if we start with $k = 6$ alleles, this means that the average number of generations back until there is a coalescence of six alleles into five alleles is, from equation 5.9, as

$$\frac{4N}{k(k-1)} = \frac{4N}{6(5)} = \frac{4N}{30} = \frac{2N}{15} \text{ generations}$$

If $N = 100$, this means that the time back to the first coalescent event is $\frac{200}{15} = 13.3$ generations (as will be clear shortly, it is convenient to express generations in terms of $2N$). At this point we have $k = 5$ alleles, and the average number of generations back until these five alleles coalesce into four alleles is given by equation (5.9) as

$$\frac{4N}{k(k-1)} = \frac{4N}{5(4)} = \frac{4N}{20} = \frac{2N}{10} \text{ generations}$$

Equation (5.9) can be used for earlier coalescent events. Extending back to three alleles would take $2N/6$ generations on average. From there, it will take $2N/3$ generations to coalesce into two alleles, and $2N$ generations for the final coalescence

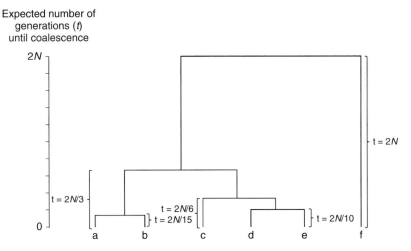

FIGURE 5.16 Expected times to coalescence. The average number of generations (t) needed for coalescent events given six alleles (labeled a through f) computed using equation (5.9). The first coalescent event (alleles a and b) reduces the number of alleles from $k = 6$ to $k = 5$ and is expected to occur on average $2N/15$ generations in the past. The next coalescent event (alleles d and e) reduces the number of alleles from $k = 5$ to $k = 4$ and is expected to occur $2N/10$ generations in the past. The process continues until all lineages coalesce to a single allele $2N$ generations in the past.

(which is why all of the above have been scaled by $2N$). Thus, the time until coalescence increases as we go back farther into the past, as shown in Figure 5.16. Note that the first few coalescent events occur relatively quickly, but it takes much longer for the last coalescent event where the number of alleles coalesces from $k = 2$ to $k = 1$.

The tree shown in Figure 5.16 reflects the genealogical relationship between the six alleles in the present day. What about their genetic relationship? If the ancestral allele at the top of the tree is passed on generation to generation without change, then all six alleles in the present day will be the same. However, mutations occur over time, such that any mutations that are unique to a particular line in the tree will be passed on to all descendants along that line (as had been shown in the example in Figure 4.6 in the previous chapter). Coalescent models that incorporate mutation and genetic drift are very useful for analyzing genetic variation. Observed levels of genetic variation can be compared to expectations generated by using simulations based on various models of population history. Methods of analysis based on coalescent theory allow estimation of the age of the most recent common ancestor. Such methods are useful in estimating the initial age of a mutation (several examples will be mentioned in Chapter 7) as well as reconstructing the genetic history of our species (discussed further in Chapter 9).

Because genetic drift (and thus coalescence) is a random process, the coalescent times shown in Figure 5.16 are expected averages, just as the expected outcome of a coin flip is a ratio of $50:50$ for heads and tails. In coalescent theory, there is also a large amount of variation expected around this average, so that any

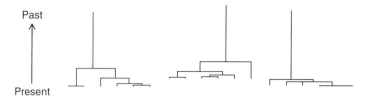

FIGURE 5.17 Simulated coalescent trees for six alleles. All trees are drawn to the same scale. Although the trees are all roughly similar, there are differences in the specific sequence of allele coalescence and the time of coalescence because coalescence is a random process. These trees were all produced using the TreeToy Java program (http://www.xmission.com/ ~wooding/TreeToy/) setting the theta parameter equal to 5—see the program documentation and Hamilton (2009:274) for details.

given coalescent tree will be different, in terms of which alleles coalesce as well as when they coalesce. This random nature is depicted in Figure 5.17, which shows three simulations of coalescence of six alleles under the assumption of constant population size. Even with just three simulations, you can see that different alleles coalesce at different times. If you would like to simulate more coalescent trees, consult the references given in the figure legend to use a web-based coalescent simulation program.

B. Coalescent Theory and Demographic History

Coalescent trees can be generated using a variety of different models and assumptions, which can provide us with considerable insight into possible evolutionary outcomes. One particularly useful feature of coalescent theory is that we can examine different demographic histories. For example, what happens to a coalescent tree if there is a major change in population size, such as rapid population growth? As an example, Figure 5.18 shows three coalescent trees generated with the same

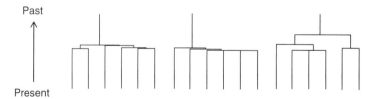

FIGURE 5.18 Simulated coalescent trees for six alleles under a demographic history of rapid population growth. All trees are drawn to the same scale (and to the same scale as in Figure 5.17). The simulation increased the size of the population 1000-fold halfway along the scale from present to past to approximate rapid exponential growth (Rogers and Harpending 1992). The shapes of these trees are different when compared to the simulations in Figure 5.17, which were based on constant population size. The trees here are characteristically star-shaped (or comb-shaped), with most coalescent events occurring around the time of rapid population growth. These trees were all produced using the TreeToy Java program (http://www.xmission.com/~wooding/TreeToy/) setting the theta parameter equal to 5, the growth value equal to 1000, and the tau parameter equal to 15—see the program documentation and Hamilton (2009:274) for details.

parameters as in Figure 5.17 but with a different history of population size. Here, the simulation program was set up to increase dramatically in size halfway from the past to the present. The results (using the same scale as in Figure 5.17) show a much different pattern. Under rapid population growth, most of the coalescent events occur at about the same time, corresponding closely to the time of major population growth. Coalescent trees such as those in Figure 5.18 are often referred to as *star-shaped* because the different lineages are all of similar length radiating from a common origin, much in the same way as light rays radiate out from a star. The star analogy is seen more clearly if we arrange the lineages of each figure in a circle around the last coalescent event; many prefer the term *comb-shaped* because the schematic images resemble a hair comb.

The reason for this star (or comb) pattern is clear when we look at population size backward in time from the present to the past (i.e., retrospectively)—present-day population size is large than in the past. In larger populations, the probability of coalescence is less than in a small population. As we trace the tree backward in time to the point of population expansion, we reach a point when population size was smaller and coalescence was more likely. It is at this point that most of the coalescing occurs.

When we compare Figures 5.17 and 5.18, we see that different demographic histories will result in different coalescent trees. We can then use these predictions for comparison with various measures of genetic diversity to determine what happened in the past [see Rogers and Harpending (1992) and Jobling et al. (2004) for examples]. For example, star-shaped gene trees tell us that there has been a lot of population growth in the past, such as a recovery from a bottleneck.

VI. SUMMARY

There is a certain amount of randomness in the universe, and the evolutionary process is no exception. Whenever a parent has a child, only half of the parent's nuclear DNA is passed on to the child, and whatever gets passed on to one child might differ from what gets passed on to the next child. If you have a heterozygous genotype, such as Aa, you expect to pass on the A allele half the time and the a allele half the time. These are average expectations, and if you have four children, you might easily pass on the A allele to all four. Genetic drift can be considered as the same process at the level of the entire population. Even if there are no mutations or selection, it is unlikely that the allele frequencies in the offspring will be identical to the parent population.

Genetic drift can lead the allele frequency in a population to increase, decrease, or stay the same. Because drift is a random process, there is no inherent direction to this change. Genetic drift will continue, generation after generation, until an allele is either lost (extinction) or replaces all other alleles (fixation). The probability of extinction or fixation depends on the initial allele frequency. A new mutant allele will have a very low initial frequency and consequently will usually be lost through genetic drift. However, in some small amount of the time, genetic drift will actually wind up leading to fixation of the mutant allele.

Genetic drift occurs in all populations, but the smaller the population, the greater its effect. Large fluctuations in allele frequency are more likely in small populations. *Population size* refers here to the breeding population size, the number of adults of reproductive age in the population. In simple models of genetic drift, we make a number of assumptions about the demography of the population that are sometimes violated in the real world. In such cases, we refer to the concept of an effective population size that is an adjustment to the breeding population size to give an estimate of the actual genetic size of the population. For example, if a population grew rapidly in size, the present-day population size would be an overestimate (and thus an underestimate of drift). Instead, we use an adjustment to get an average population size that would produce the same level of genetic drift under a model of constant size that we would see when the population changes in size. There are numerous different types of adjustment for effective population size, some of which have little impact on human populations, and some of which have greater significance. Of particular anthropological interest is how demography and cultural variation can influence effective population size and the level of genetic drift.

When we look at the process of genetic drift, we look forward from the past to the present to see what happens to different alleles over time, including which ones become extinct. A branch of population genetics known as *coalescent theory* looks at drift in reverse, by tracking alleles back in time to see which ones have a common ancestor in a previous generation. Over time, all alleles in the present day coalesce into fewer and fewer ancestral alleles, until ultimately there is a single ancestral allele. Coalescent theory allows many valuable inferences regarding this most recent common ancestor (MRCA) as well as demographic history.

APPENDIX 5.1 DECAY OF HETEROZYGOSITY OVER TIME DUE TO GENETIC DRIFT

To show how drift reduces genetic variation (heterozygosity) over time, we return to the concept of identity by descent from Chapter 3. In a population of $2N$ alleles, the probability that someone will have inherited any specific allele is $1/2N$. The probability of inheriting the same allele twice is $(1/2N)(1/2N)$. This probability applies to that specific allele. If we extend this to consider *any* of the $2N$ alleles, then the probability of someone having the same allele because of common descent is derived using the OR rule over all possible alleles, giving

$$(1/2N)(1/2N)2N = 1/2N \tag{A5.1}$$

Because this is the probability that two alleles are identical by descent, the probability that two alleles are not inherited from a common ancestor in the previous generation is 1 minus equation (A5.1):

$$\left(1 - \frac{1}{2N}\right) \tag{A5.2}$$

For any generation t, these two equations refer to probabilities of identity by descent in the previous generation, $t - 1$. If we are looking at the total probability of two alleles being identical due to descent, equation (A5.1) does not suffice because there is a possibility that the proportion of the population represented by equation (A5.2) might be identical by descent in *a previous generation*. If we define this probability as f_{t-1}, we can now deal with two ways in which two alleles can be identical by descent in generation t. The first is equation (A5.1), which refers to the previous generation. The second possibility is the probability that two alleles are not identical by descent one generation earlier [probability $= (1 - 1/2N)$] AND are identical by descent from a previous generation (probability $= f_{t-1}$). The overall probability of this second case is determined by using the AND rule, so these last two terms are multiplied, giving

$$\left(1 - \frac{1}{2N}\right)f_{t-1} \tag{A5.3}$$

We now have two different ways of ensuring that a pair of alleles can be identical by descent, and we look at the total probability of one *or* the other happening. Using the OR rule, we add the two probabilities [equations (A5.1) and (A5.3)] to obtain an overall probability of identity by descent in generation t as

$$f_t = \frac{1}{2N} + \left(1 - \frac{1}{2N}\right)f_{t-1} \tag{A5.4}$$

Note that we use the lowercase symbol f to refer to the probability of descent due to finite population size in order to distinguish it from the probability of descent due to inbreeding, F, used in Chapter 3.

We can now relate equation (A5.4) to homozygosity by noting that the probability of heterozygosity can be defined relative to homozygosity as $H = 1 - f$. We can express equation (A5.4) in terms of heterozygosity by subtracting both sides of equation (A5.4) from 1, giving

$$1 - f_t = 1 - \left[\frac{1}{2N} + \left(1 - \frac{1}{2N}\right)f_{t-1}\right]$$

Expanding all the terms gives

$$1 - f_t = 1 - \frac{1}{2N} - \left(1 - \frac{1}{2N}\right)f_{t-1}$$

which, in turn, gives

$$1 - f_t = \left(1 - \frac{1}{2N}\right)(1 - f_{t-1}) \tag{A5.5}$$

Because $H = 1 - f$, and keeping track of the generational subscripts, equation (A5.5) can be rewritten as

$$H_t = H_{t-1}\left(1 - \frac{1}{2N}\right)$$

If we start with an initial value of heterozygosity of H_0 for generation $t = 0$, the expected heterozygosity in generation $t = 1$ is

$$H_1 = H_0 \left(1 - \frac{1}{2N} \right)$$

We can then give the expected heterozygosity in generation $t = 2$ as

$$H_2 = H_1 \left(1 - \frac{1}{2N} \right)$$

and then substitute the expected value for H_1, giving us

$$H_2 = H_0 \left(1 - \frac{1}{2N} \right) \left(1 - \frac{1}{2N} \right) = H_0 \left(1 - \frac{1}{2N} \right)^2$$

If we repeat the same process for generation $t = 3$, we get

$$H_3 = H_0 \left(1 - \frac{1}{2N} \right)^3$$

In general

$$H_t = H_0 \left(1 - \frac{1}{2N} \right)^t$$

which is equation (5.7) in the text.

APPENDIX 5.2 EXPECTED HETEROZYGOSITY AT EQUILIBRIUM IN THE INFINITE ALLELES MODEL

Here, we look at the interaction between mutation (which increases variation) and genetic drift (which reduces variation) in a population. We start with equation (A5.4), which predicts identity by descent as a function of genetic drift. Under mutation, identity by descent will be possible only if a given allele has *not* mutated. By definition, the mutation rate μ is the probability that an allele will mutate, which means that the probability of an allele *not* mutating is $(1 - \mu)$. Therefore, the probability of having *two* alleles, neither of which has mutated, is $(1 - \mu)(1 - \mu) = (1 - \mu)^2$. We now modify equation (A5.4) to consider the probability of identity by descent due to drift *and* neither allele mutating, we get (using the AND rule)

$$f_t = (1 - \mu)^2 \left[\frac{1}{2N} + \left(1 - \frac{1}{2N} \right) f_{t-1} \right] \qquad (A5.6)$$

This equation can be difficult to work with, but population geneticists working with equations with mutation frequently use approximations that are simpler, yet very accurate. For example, when we expand the term $(1 - \mu)^2$, we obtain $1 - 2\mu + \mu^2$. Because the mutation rate (μ) is very small, this means that μ^2 will be very, very small, and for all practical purposes equal to zero. This also means that the expression $(1 - \mu)^2$ is approximately equal to the much simpler expression $(1 - 2\mu)$. If we substitute this approximation back into equation (A5.6) and multiply out all of the terms, we get

$$f_t \approx \frac{1 - 2\mu}{2N} + (1 - 2\mu)\left(1 - \frac{1}{2N}\right)f_{t-1}$$

$$f_t \approx \frac{1}{2N} - \frac{\mu}{N} + \left(1 - \frac{1}{2N} - 2\mu + \frac{\mu}{N}\right)f_{t-1}$$

(where the symbol $\approx$ means "approximately equal to"). At this point, we can simplify things further by noting that the expression μ/N, which appears twice in the above equation, is also essential equal to zero (because the mutation rate is so small and population size is so much larger, the ratio is practically zero in all cases). By setting $\mu/N = 0$, we now obtain

$$f_t \approx \frac{1}{2N} + \left(1 - \frac{1}{2N} - 2\mu\right)f_{t-1} \tag{A5.7}$$

Over time, the variation that is lost as a result of genetic drift is offset by variation added by mutation. Eventually, these two forces balance each other, as in the flowing water analogy of Figure 5.13. When this equilibrium is reached, there will be no further change in f, such that

$$f = f_t = f_{t-1}$$

We can solve for this equilibrium by setting f_t and f_{t-1} in equation (A5.7) both equal to f, giving

$$f \approx \frac{1}{2N} + \left(1 - \frac{1}{2N} - 2\mu\right)f$$

$$f - \left(1 - \frac{1}{2N} - 2\mu\right)f \approx \frac{1}{2N}$$

$$f - \left(f - \frac{f}{2N} - 2\mu f\right) \approx \frac{1}{2N}$$

$$f - f + \frac{f}{2N} + 2\mu f \approx \frac{1}{2N}$$

$$\frac{f}{2N} + 2\mu f \approx \frac{1}{2N}$$

$$f\left(\frac{1}{2N} + 2\mu\right) \approx \frac{1}{2N}$$

This equation can be further simplified by multiplying 2μ by $(2N/2N)$ to obtain

$$f\left(\frac{1+4N\mu}{2N}\right) \approx \frac{1}{2N}$$

which further simplifies to

$$f \approx \frac{1}{1+4N\mu} \tag{A5.8}$$

Recalling from equation (A5.1) that heterozygosity can be expressed as $H = 1 - f$, and applying this to equation (A5.8), gives

$$H = 1 - f \approx 1 - \frac{1}{1+4N\mu}$$

$$H \approx \frac{1+4N\mu}{1+4N\mu} - \frac{1}{1+4N\mu}$$

$$H \approx \frac{4N\mu}{1+4N\mu}$$

which is equation (5.8) in the text.

APPENDIX 5.3 COMPUTATION OF NUCLEOTIDE DIVERSITY

Genetic variation can be assessed at the molecular level with a number of different measures, including looking at the proportion of nucleotide differences among a set of DNA sequences. In order to illustrate this method, consider the following set of four DNA sequences, each consisting of 10 sites (in reality, these numbers are too small to be useful, but this example is for illustration so that this restriction is not a problem):

Sequence 1: C C C A T T C A T C
Sequence 2: C C G A T T C A T C
Sequence 3: C C C A T T C T T C
Sequence 4: C C C A T T C T T C

We now want to compare each sequence to all other sequences looking for the number of **mismatches**. For example, when comparing sequences 1 and 2, we see that there is a mismatch in the third position in the sequences—sequence 1 has the nucleotide C, and sequence 2 has the nucleotide G. Thus, we have one mismatch in 10 nucleotides, or $\frac{1}{10} = 0.1$ differences per site. We denote this average difference d. We then compute d between all possible pairs of the four sequences, which gives the following six values:

Sequences 1 and 2, $d = \frac{1}{10} = 0.1$
Sequences 1 and 3, $d = \frac{1}{10} = 0.1$

Sequences 1 and 4, $d = \frac{1}{10} = 0.1$

Sequences 2 and 3, $d = \frac{2}{10} = 0.2$

Sequences 2 and 4, $d = \frac{2}{10} = 0.2$

Sequences 3 and 4, $d = \frac{0}{10} = 0.0$

We now compute the average value of d over these six comparisons, giving a nucleotide diversity of $\pi = (0.1 + 0.1 + 0.1 + 0.2 + 0.2 + 0.0)/6 = \frac{0.7}{6} = 0.1167$.

In general terms, the computational formula is

$$\pi = \frac{\sum_{i<j}^{m} d_{ij}}{c}$$

where d_{ij} refers to the proportion of nucleotide mismatches between sequences i and j, and c is the total number of comparisons, $c = m(m-1)/2$, where m is the number of sequences ($m = 4$ in the above example, so $c = 6$) (Nei and Kumar 2000). The notation below the summation sign ($i < j$) is an instruction to count i from 1 to m and j from 1 to m, but only for those combinations where $i < j$ (this prevents us from comparing a sequence to itself, or comparing the same two sequences more than once; for example, comparing sequences 2 and 1 is the same as comparing sequences 1 and 2). In the above example, the values of i and j would be 1–2, 1–3, 1–4, 2–3, 2–4, and 3–4.

CHAPTER 6

MODELS OF NATURAL SELECTION

The phrase "survival of the fittest," first coined by philosopher and sociologist Herbert Spencer, was later used by Charles Darwin to refer to the process of natural selection. In common usage, the phrase tends to convey a stilted view of natural selection, whereby the "fittest" are often considered as those individuals who are strongest, fastest, and smartest and thus win in a competitive battle with others. The problem with this image is that we too often equate the evolutionary concept of fitness with the more general attributes of physical fitness. Although size and strength can increase one's probability of survival in some cases, this is not always the case, and there is no necessary connection between this image of physical fitness and evolutionary fitness. In some environments, for example, it might be more adaptive to be *smaller* because of the advantages of smaller body size, such as the need for less food. What matters in natural selection is the *net* advantage of a given trait or traits, which, in turn, can vary by environment. In some cases, larger body size might have the net advantage, and other cases a smaller body size might have this net advantage.

Another problem with confusing "survival of the fittest" with the general process of natural selection is that the emphasis is on survival, which implies that only the differential survival of individuals matters. In an evolutionary context, what matters is the survival of an individual's genes, which means that natural selection must also take into consideration differences in fertility. Natural selection involves differences in survival *and* reproduction. In order to contribute genetically to the next generation, an individual must survive until reproduction and then reproduce. Fitness must be regarded in this general way and not confined to any narrow definition of "fitness."

This chapter begins with a mathematical consideration of fitness using basic concepts of probability. A simple example is then used to show how differences in fitness between genotypes lead to a change in allele frequencies. This example is used to construct a general model of selection for a locus with two alleles, which is then used to demonstrate several different forms of selection. Chapter 7 continues

Human Population Genetics, First Edition. John H. Relethford.
© 2012 Wiley-Blackwell. Published 2012 by John Wiley & Sons, Inc.

the discussion of natural selection by providing a number of case studies that illustrate natural selection in human populations, past and present.

I. HOW DOES NATURAL SELECTION WORK?

A common approach to natural selection in biology classes is to simulate the process using laboratory organisms such as fruit flies, whose short generation lengths make them amenable to observing evolution over short periods of time. Natural selection can also be modeled using simple mathematics, an approach that provides many insights into how natural selection can operate in nature.

A. Absolute and Relative Fitness

As noted above, the term **fitness** goes beyond any simple equation with physical fitness. When modeling natural selection, fitness simply refers to the probability of survival and reproduction. Models of natural selection are based on each genotype being associated with a specific fitness, which expresses the relative probability of representation in the next generation. This might sound rather abstract, but a simple example shows how this works.

We start with our standard model of a locus with two alleles, A and a, such that we have three genotypes: AA, Aa, and aa. Imagine that we have the following numbers of individuals in a population at birth: 500 individuals with genotype AA, 1000 with genotype Aa, and 500 with genotype aa, for a total of 2000 individuals. Imagine that we examine this population later in time to see how many have survived to reproductive age, and find that there are now 1575 individuals made up of 450 with genotype AA, 900 with genotype Aa, and 225 with genotype aa. We now compare the number of each genotype for the 2 times in the following table:

	AA	Aa	aa
At birth	500	1000	500
At adulthood	450	900	225

We can see that there has been some mortality; of the original 2000 individuals, 1575 survived to adulthood and 425 did not. However, if we look more closely, we see that this mortality was not the same for each genotype. Our interest is in the proportional change for each genotype, which is known as **absolute fitness**, which can be computed for each genotype by taking the ratio of the numbers at two points in time, which gives

AA	Aa	aa
$\frac{450}{500} = 0.9$	$\frac{900}{1000} = 0.9$	$\frac{225}{500} = 0.45$

We can now easily see that although some individuals from each genotype did not survive, this mortality was not the same for all genotypes. Proportionally, far more individuals died that had genotype aa than was the case for genotypes AA or Aa. In this example, we considered only differential mortality, but we can also consider similar effects from differential fertility, where there are differences in the number of births by genotype.

Mathematically, it is easier to deal with a quantity known as **relative fitness**, which expresses fitness relative to the most fit genotype. Relative fitness is typically denoted by the letter w, with subscripts used to refer to the different genotypes. Thus, the symbol w_{AA} is used to refer to the relative fitness of genotype AA, w_{Aa} is used to refer to the relative fitness of genotype Aa, and w_{aa} is used to refer to the relative fitness of genotype aa. These values are easily computed by dividing each absolute fitness value by the highest absolute fitness value, which sets the highest relative fitness equal to 1 by definition. In terms of the example above, the highest absolute fitness, shared by both genotypes AA and Aa, is 0.9, which means that the **relative fitness** values are

AA	Aa	aa
$w_{AA} = \dfrac{0.9}{0.9} = 1.0$	$w_{Aa} = \dfrac{0.9}{0.9} = 1.0$	$w_{aa} = \dfrac{0.45}{0.9} = 0.5$

In relative terms, these numbers mean that for every 100 individuals with genotype AA (or Aa) that survive, only 50 with genotype aa survive. This type of variation in fitness is typical of a recessive allele that is harmful when two copies are inherited (the homozygous genotype aa), but not when only one copy is inherited (the heterozygous genotype Aa).

If the absolute fitness of all three genotypes were the same, then all of the relative fitness values would also be the same, and all would equal 1.0. In this case, there would be no differential survival and/or reproduction by genotype, and allele frequencies would remain at Hardy–Weinberg equilibrium. If at least one fitness value differed among the three genotypes, then natural selection would occur. Natural selection can sometimes cause an allele to increase in frequency, sometimes decrease in frequency, and sometimes reach a balance. The specific effect of natural selection depends on the fitness of each specific genotype. Several examples of likely forms of natural selection will be described later in this chapter.

B. A Simulation of Natural Selection

At this point, we will relate the concept of relative fitness to natural selection by means of a simple mathematical simulation that uses some of the basic rules of probability outlined in Chapter 1. This simulation will use the standard model of a locus with two alleles, A and a. In order to simulate the process, we need the initial allele frequencies (p and q) and the relative fitness of each genotype: w_{AA}, w_{Aa}, and w_{aa}. For this example, we will start with allele frequencies of $p = 0.5$ and $q = 0.5$ and relative fitness values of: $w_{AA} = 1.0$, $w_{Aa} = 1.0$, and $w_{aa} = 0.5$.

Step 1: Start with Genotype Frequencies under Hardy–Weinberg Equilibrium

We start off by determining the expected genotype frequencies of the population *before selection* by deriving the Hardy–Weinberg proportions. Given $p = 0.5$ and $q = 0.5$, these are

AA	Aa	aa
$p^2 = 0.25$	$2pq = 0.50$	$q^2 = 0.25$

Step 2: Compute Change in Genotype Frequencies Due to Selection

If there were no selection, the proportions above would also be the expected proportions in the next generation. Under selection, however, we need to model differential fitness. First, we list the fitness values under their respective genotypes:

	AA	*Aa*	*aa*
Before selection	0.25	0.50	0.25
Fitness	$w_{AA} = 1.0$	$w_{Aa} = 1.0$	$w_{aa} = 0.5$

Each row here represents probabilities associated with each genotype. The first row (before selection) shows the probabilities of having a given genotype. The second row (fitness) shows the probabilities that someone with a given genotype will contribute genetically to the next generation. Therefore, the probability that someone has a given genotype AND contributes genetically to the next generation is solved using the AND rule, which entails multiplying these probabilities. The expected genotype proportions *after* selection are therefore obtained as follows:

	AA	*Aa*	*aa*
Before selection	0.25	0.50	0.25
Fitness	1.0	1.0	0.5
After selection	$0.25 \times 1.0 = 0.250$	$0.50 \times 1.0 = 0.500$	$0.25 \times 0.5 = 0.125$

An important quantity is the sum of the genotype frequencies after selection:

$$\overline{w} = 0.250 + 0.500 + 0.125 = 0.875$$

This number is the **mean fitness** across the three genotypes (denoted $\overline{w}$). The mean fitness is not simply the average of the three relative fitness values, which would be $= [(1 + 1 + 0.5)/3]$, but instead is a mean where each relative fitness is weighted by the frequency of that genotype in the population after selection. The mean fitness tells us how much selection has taken place relative to the case where no selection has occurred (where the mean fitness under Hardy–Weinberg equilibrium would be equal to 1.0). Here, the mean fitness of 0.875 means that 87.5% of the individuals survive, and therefore that $100 - 87.5 = 12.5\%$ did not survive.

Step 3: Normalize Genotype Frequencies

What effect do you think selection has had on the allele frequencies? It may seem intuitive that the frequency of the a allele will decrease because we are selecting out half of those individuals with the aa genotype. This is correct, but to see the exact effect, we need to compute the allele frequencies after selection. Although we usually compute allele frequencies from genotype frequencies using equation (2.3), where we add the frequency of one homozygote to half the frequency of the heterozygote, we cannot do so in this case because the genotype frequencies after selection do not add up to 1.0. Instead, they add up to the mean fitness. To circumvent this problem, we need to express the genotype frequencies after selection relative to the mean fitness, which is easily done by dividing each genotype frequency by the mean fitness. This process is called *normalization*. After doing this, our table has a new row and looks like this:

	AA	Aa	aa
Before selection	0.25	0.50	0.25
Fitness	1.0	1.0	0.5
After selection	0.250	0.500	0.125
Normalized genotype frequencies	$\dfrac{0.250}{0.875} = 0.2857$	$\dfrac{0.500}{0.875} = 0.5714$	$\dfrac{0.125}{0.875} = 0.1429$

Note that the sum of the normalized genotype frequencies now adds up to 1.0 $(0.2857 + 0.5714 + 0.1429 = 1.0)$.

Step 4: Compute New Allele Frequencies

Given the normalized genotype frequencies after selection, we can now compute the new allele frequencies using equation (2.3):

$$p = f_{AA} + \frac{f_{Aa}}{2} = 0.2857 + \frac{0.5714}{2} = 0.5714$$

$$q = \frac{f_{Aa}}{2} + f_{aa} = \frac{0.5714}{2} + 0.1429 = 0.4286$$

We can easily see the impact of natural selection after a single generation. Here, the frequency of the a allele has decreased as expected, from $q = 0.5$ to 0.4286. Because the frequency of q decreases, the frequency of p increases, from 0.5 to 0.5714.

When we begin to track changes in allele frequencies due to selection over many generations, it will be useful to keep track of the exact amount of change from one generation to the next. If we refer to p and q as the allele frequencies before selection and the symbols p' and q' as the allele frequencies after selection, then the amount of change in a given generation (Δ) will be equal to

$$\Delta p = p' - p = 0.5714 - 0.5 = 0.0714$$

$$\Delta q = q' - q = 0.4286 - 0.5 = -0.0714$$

The value for Δp is positive because the A allele increased in frequency, and the value for Δq is negative because the a allele decreased in frequency.

Step 5: Extend the Results to the Next Generation

We can now repeat steps 1–4 by taking the new allele frequencies ($p = 0.5714$ and $q = 0.4286$) and starting over and the assuming that the fitness values remain the same. Filling in all the rows of the table that we built, we now obtain

	AA	Aa	aa
Before selection	0.3265	0.4898	0.1837
Fitness	1.0	1.0	0.5
After selection	0.3265	0.4898	0.0919
Normalized genotype frequencies	0.3595	0.5393	0.1012

Finally, after selection, the new values of p and q after the second generation of selection are $p = 0.6292$ and $q = 0.3708$, respectively. The frequency of the a allele has continued to decrease, and Δq now equals $0.3708 - 0.4286 = -0.0578$. The mean fitness (the sum of the unnormalized genotype frequencies after selection) has increased from 0.875 in the first generation to

$$\overline{w} = 0.3265 + 0.4898 + 0.0919 = 0.9082$$

in the second generation. This increase is due to the reduction in individuals susceptible to selection (those with genotype aa in this particular example).

The above process can be repeated generation after generation. Although the computations could be done by hand, they would become rather tiresome quickly, and it is much easier to use a spreadsheet program. Using such a program, I simulated 100 generations of selection. The results are shown in Figures 6.1–6.3. Figure 6.1 shows how the frequency of the a allele (q) decreases over time because of selection against individuals with genotype aa. This decrease is rapid at first, and then slows down because there are proportionately fewer individuals with genotype aa to select against each generation.

Figure 6.2 shows the amount of allele frequency change per generation (Δq). The amount of change is negative from one generation to the next, because the frequency of q decreases over time. However, the magnitude of this change decreases over time as the impact of selection declines with fewer aa individuals. Thus, Δq approaches a value of zero, which is at equilibrium when $q = 0$ and there will be no further change. In a mathematical sense, q will never reach 0 but will approach it. In a practical sense, there will come a point where q is so low that it becomes zero because of genetic drift.

Figure 6.3 shows how mean fitness increases over time to approach a theoretical maximum where $p = 1, q = 0$, and the mean fitness $= 1$. This graph is important in understanding natural selection, because it is a process that leads, generation after generation, toward maximizing mean fitness.

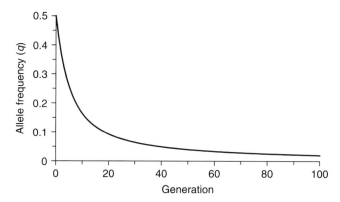

FIGURE 6.1 Allele frequency under selection against the recessive homozygote. Initial allele frequencies are $p = 0.5$ and $q = 0.5$. Fitness values are $w_{AA} = 1.0$, $w_{Aa} = 1.0$, and $w_{aa} = 0.5$. The frequency of the a allele (q) decreases over time; the decrease is rapid at first and then slows down. Over time, the frequency of q will approach zero.

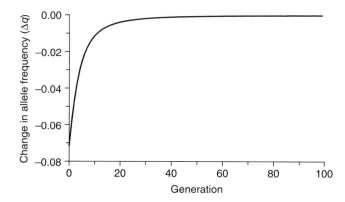

FIGURE 6.2 Allele frequency change per generation under selection against the recessive homozygote. Initial allele frequencies are $p = 0.5$ and $q = 0.5$. Fitness values are $w_{AA} = 1.0$, $w_{Aa} = 1.0$, and $w_{aa} = 0.5$. With each generation the amount of change is negative because the allele frequency q is decreasing over time (see Figure 6.1). Over time, the amount of change approaches zero.

II. A GENERAL MODEL OF NATURAL SELECTION

Now that we have seen an example of natural selection over time, we can apply the methods used to simulate natural selection to develop a general mathematical model of natural selection that can be used for any conceivable set of fitness values. Thus, we will be able to understand easily not only selection against a recessive allele but also selection against a dominant allele, selection against the heterozygote, and selection for the heterozygote, among other possible scenarios.

In order to simulate any case of selection for a single locus with two alleles, we need the initial allele frequencies and the relative fitness for each genotype. We then can take the tables presented in the five-step example above and substitute the

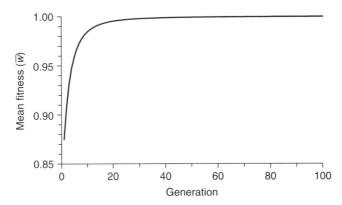

FIGURE 6.3 Mean fitness in each generation under selection against the recessive homozygote. Initial allele frequencies are $p = 0.5$ and $q = 0.5$. Fitness values are $w_{AA} = 1.0$, $w_{Aa} = 1.0$, and $w_{aa} = 0.5$. Mean fitness increases with each generation, rapidly at first and then slowing down to approach a maximum value of 1.

TABLE 6.1 A General Model of Natural Selection[a]

Time Frame or Parameter	Genotype		
	AA	*Aa*	*aa*
Before selection	p^2	$2pq$	q^2
Fitness	w_{AA}	w_{Aa}	w_{aa}
After selection	$p^2 w_{AA}$	$2pqw_{Aa}$	$q^2 w_{aa}$
Normalized genotype frequencies	$(p^2 w_{AA})/\overline{w}$	$(2pqw_{Aa})/\overline{w}$	$(q^2 w_{aa})/\overline{w}$

[a]This table presents the values used to simulate natural selection for a single-locus model with two alleles, A and a. The model requires the initial allele frequencies p and q and the fitness values for each genotype. The genotype frequencies before selection are the Hardy–Weinberg equilibrium values. The genotype frequencies after selection are obtained by multiplying the frequencies before selection by their respective fitness values. The normalized genotype frequencies are obtained by dividing the frequencies after selection by the mean fitness, designated by the symbol $\overline{w}$ (the bar above the letter is standard shorthand for the mean, which in this case is the mean value of w, which is fitness). Here, $\overline{w} = p^2 w_{AA} + 2pqw_{Aa} + q^2 w_{aa}$.

appropriate mathematical variables for the specific values used in the example. Table 6.1 shows the expected genotype frequencies before selection (obtained from Hardy–Weinberg equilibrium), the genotype frequencies after selection, and the normalized genotype frequencies after selection. The formulas presented in Table 6.1 are used for deriving several quantities below. The derivations presented in this chapter are taken from Ayala (1982).

The first quantity that we derive (and that is needed to compute the normalized genotype frequencies) is the mean fitness, computed as the sum of the genotype frequencies after selection:

$$\overline{w} = p^2 w_{AA} + 2pqw_{Aa} + q^2 w_{aa} \qquad (6.1)$$

We now derive the allele frequencies after a single generation of selection, labeled as p' and q', by using equation (2.3) on the appropriate normalized genotype frequencies after selection from Table 6.1. For p', this gives

$$p' = f_{AA} + \frac{f_{Aa}}{2}$$

from equation (2.3), which, when used with the values in Table 6.1, gives

$$p' = \frac{p^2 w_{AA}}{\overline{w}} + \frac{2pq w_{Aa}}{2\overline{w}} = \frac{p^2 w_{AA} + pq w_{Aa}}{\overline{w}}$$

We then factor out p to get

$$p' = \frac{p(p w_{AA} + q w_{Aa})}{\overline{w}} \qquad (6.2)$$

We do the same thing for q' to obtain

$$q' = \frac{q(p w_{Aa} + q w_{aa})}{\overline{w}} \qquad (6.3)$$

(*Note*: To follow this, be sure to keep the subscripts for the fitness values straight!) We could then solve for additional generations by setting the initial allele frequencies equal to p' and q' and solving for the next generation (again, assuming that the fitness values do not change).

We are also interested in the amount of change in allele frequencies: Δp and Δq. Derivation of these values is a bit more involved algebraically, and is presented in Appendix 6.1. These values are

$$\Delta p = \frac{pq[p(w_{AA} - w_{Aa}) + q(w_{Aa} + w_{aa}]}{\overline{w}} \qquad (6.4)$$

and

$$\Delta q = \frac{pq[p(w_{Aa} - w_{AA}) + q(w_{aa} - w_{Aa})]}{\overline{w}} \qquad (6.5)$$

III. TYPES OF NATURAL SELECTION

Given the general model above, we will use equations (6.1)–(6.5) to explore several common types of natural selection.

A. Selection against the Recessive Homozygote

The worked example of natural selection given earlier (Figures 6.1–6.3) involved selection against the recessive homozygote. This type of selection is common with a number of genetic disorders where having two copies of a harmful recessive allele lowers an individual's fitness. In the most extreme case, that of a lethal recessive, the fitness is zero. The general model of natural selection can be used to explore some general implications of selection against the recessive homozygote.

Here, we use our standard model of a locus with two alleles, and let A represent the dominant allele and a represent the recessive allele.

Models of natural selection are best understood by using a measure known as the **selection coefficient**, which is the opposite of fitness. As fitness represents the probability of survival and reproduction, the selection coefficient is the probability of *not* surviving and reproducing. We use w to represent fitness and s to represent the selection coefficient. The two measures are related as $w + s = 1$. As an example, values of $w = 0.8$ and $s = 0.2$ mean that 80% survive and 20% are selected against. The models for natural selection work out more easily mathematically if we express the fitness in terms of the selection coefficient as $w = 1 - s$.

To consider the effect of selection against the recessive homozygote, we need to assign a relative fitness for the aa genotype that is less than those for the other genotypes (which are assigned a fitness of 1). In this case, we consider s as the selection coefficient associated with the recessive homozygote. We thus set the fitness values for this type of selection as

$$w_{AA} = 1$$

$$w_{Aa} = 1$$

$$w_{aa} = 1 - s$$

In this case, the fitness of the recessive homozygote is reduced by the quantity s relative to the other two genotypes. We now plug these values into equations (6.1)–(6.5) to obtain equations that describe the effect of selection against recessive homozygotes. The mean fitness, obtained from equation (6.1), is

$$\overline{w} = p^2 w_{AA} + 2pq w_{Aa} + q^2 w_{aa}$$
$$= p^2(1) + 2pq(1) + q^2(1 - s)$$

which gives

$$\overline{w} = 1 - sq^2 \tag{6.6}$$

(see Appendix 6.2 for the complete derivation). We can see that the higher the frequency of the a allele (q) and/or the higher the amount of selection against the recessive homozygote (s), the lower the mean fitness. Mean fitness will be highest when $q = 0$ (i.e., the harmful allele has been eliminated from the population).

We can see the impact of selection on the frequency of the a allele by plugging in the fitness values above into equation (6.3), which gives

$$q' = \frac{q(p w_{Aa} + q w_{aa})}{\overline{w}}$$
$$= \frac{q[p(1) + q(1 - s)]}{\overline{w}}$$

which, following some algebraic manipulation (see Appendix 6.2), gives

$$q' = \frac{q - sq^2}{1 - sq^2} \tag{6.7}$$

This formula can be used in a spreadsheet to calculate the frequency of the a allele from one generation to the next by taking the value of q' each generation and plugging it back into equation (6.7) to get the value of q' for the next generation. For example, if $s = 0.5$ and the initial value of q is set to 0.5, we obtain

$$q' = \frac{0.5 - (0.5)(0.5)^2}{1 - (0.5)(0.5)^2} = \frac{0.5 - 0.125}{1 - 0.125} = 0.4286$$

in the next generation. To extend this another generation, we would plug this new value of q back into equation (6.7) and obtain

$$q' = \frac{0.4286 - (0.5)(0.4286)^2}{1 - (0.5)(0.4286)^2} = \frac{0.3368}{1 - 0.0918} = 0.3708$$

This process could be repeated any number of times to compute the expected value from natural selection after a given number of generations (as in Figure 6.1).

Another useful parameter is the amount of allele frequency change per generation Δq. Here, we take the fitness values from above ($w_{AA} = 1, w_{Aa} = 1, w_{aa} = 1 - s$) and substitute them into equation (6.5), giving

$$\Delta q = \frac{pq[p(w_{Aa} - w_{AA}) + q(w_{aa} - w_{Aa})]}{\overline{w}}$$

$$= \frac{pq[p(1 - 1) + q(1 - s - 1)]}{\overline{w}}$$

which gives (see Appendix 6.2)

$$\Delta q = \frac{-spq^2}{1 - sq^2} \tag{6.8}$$

Much of the behavior of selection against the recessive homozygote can be inferred directly from equation (6.8). For one thing, we can see that the term Δq will always be negative because of the negative sign in the numerator and the fact that s, p, and q will always be nonnegative by definition. (After all, what is a negative probability?) The negative nature of Δq means that the frequency of the a allele will decrease, generation after generation. In simpler terms, selection against the recessive homozygote causes the recessive allele to decrease over time. It will never increase in frequency (at least as a result of selection—it could increase because of mutation and genetic drift).

We have already seen from Figure 6.1 that under selection against the recessive homozygote, the frequency of the a allele will continue to approach a value of zero. We can also see this by solving for the equilibrium value by setting equation(6.8) equal to zero (which is the mathematical definition of equilibrium). In this case, we solve the equation

$$\Delta q = \frac{-spq^2}{1 - sq^2} = 0$$

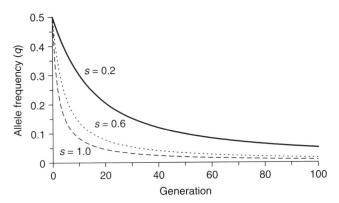

FIGURE 6.4 Selection against the recessive homozygote for different values of the selection coefficient. Initial allele frequencies are $p = 0.5$ and $q = 0.5$ in each case. Fitness values are $w_{AA} = 1, w_{Aa} = 1$, and $w_{aa} = 1 - s$. Allele frequencies were derived using equation (6.7) for three different values of the selection coefficient: $s = 0.2, s = 0.6$, and $s = 1.0$.

which means that Δq equals zero when the term $-spq^2$ is equal to zero, which could happen if s, p, or q were equal to zero. Because by definition s is not equal to zero (otherwise there would be no selection), Δq can be equal to zero only if p or q is equal to zero. However, because selection against the homozygote involves a *decrease* in the a allele (and an *increase* in the A allele), this means that an equilibrium will be reached only when q is equal to zero. As noted above, the mathematics of selection means that the frequency of a will approach—but not actually reach—zero, but the reality of the situation is that selection against the homozygote will eventually remove the a allele.

Equation (6.8) also shows the effect of the intensity of selection as measured by the selection coefficient s. For any given value of q, higher values of s will result in more allele frequency change, which means a faster approach to the final equilibrium value. This is intuitive—the greater the intensity of selection, the sooner it will occur. The effect of different values of s is shown in Figure 6.4. Higher values of s (which indicates lower fitness of the recessive homozygote) results in faster change over time. One of the curves in Figure 6.4 is particularly informative: the case of *complete* selection against the recessive homozygote. This case ($s = 0$) corresponds to a lethal recessive allele where any individual having two copies of the recessive allele will not survive and reproduce. An example in humans is Tay–Sachs disease, caused by a lethal recessive allele. Individuals that receive the lethal recessive allele from both parents are homozygous and usually die in their first few years of life. Since virtually none of the recessive homozygotes survive, does this mean that the lethal allele will be eliminated in a single generation? As seen in Figure 6.4, the answer is no, but this result might take a little bit of thought. If *all* of the recessive homozygotes die before reproducing, then how can any of the alleles continue? The answer, of course, is that *heterozygotes* carry one lethal allele and can pass it on to the next generation. Even when a recessive allele is lethal, it takes time to remove it from the population.

B. Selection against Dominant Alleles

Now that we have seen the derivation of a general model of natural selection and worked through an example in depth, we can consider different forms of natural selection. For example, what happens if the situation discussed above is reversed and we have selection *against* a dominant allele? In this scenario, the fitness of both the dominant homozygote (AA) and the heterozygote (Aa) would be reduced relative to the recessive homozygote (aa). We start by writing the fitness values for each genotype to reflect these differences:

$$w_{AA} = 1 - s$$
$$w_{Aa} = 1 - s$$
$$w_{aa} = 1$$

Mean fitness is derived by substituting these values into equation (6.1), giving

$$\overline{w} = 1 - s + sq^2 \tag{6.9}$$

(see Appendix 6.3). This equation shows us that for any value of s, mean fitness will increase as the frequency of the recessive allele (q) increases (and, therefore, as the frequency of the dominant allele decreases).

Since we are interested in the frequency of the dominant allele, we use equation (6.2) to express the allele frequency p' in the next generation, and substitute equation (6.9) for mean fitness, giving

$$p' = \frac{p(1 - s)}{1 - s + sq^2} \tag{6.10}$$

The amount of change in the dominant allele per generation is now obtained from equation (6.4), giving

$$\Delta p = \frac{-spq^2}{1 - s + sq^2} \tag{6.11}$$

(see Appendix 6.3 for derivations). This equation shows us two important facts about selection against genotypes having the dominant allele: (1) Δp will always be negative (because of the negative sign in the numerator), and the frequency of p will always decrease over time, which makes sense as it is being selected against in this example; and (2) this decrease will continue until equilibrium, defined when $\Delta p = 0$. We see that this will occur when $p = 0$. Selection against the dominant allele will eventually remove the allele from the population. In this case, removal is complete as the dominant allele is expressed in the heterozygote, and therefore cannot be retained indefinitely in the population (even in a mathematical sense). As shown in Figure 6.5, higher values of s will remove the dominant allele more quickly.

An interesting feature of selection against a dominant allele is that it is possible to remove the dominant allele in a single generation. This will happen when the

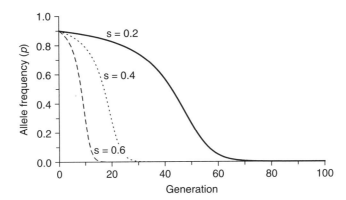

FIGURE 6.5 Selection against dominant alleles. Initial allele frequencies are $p = 0.9$ for the dominant allele (A) and $q = 0.1$ for the recessive allele (a). Fitness values are $w_{AA} = 1 - s$, $w_{Aa} = 1 - s$, and $w_{aa} = 1$. Allele frequencies were derived using equation (6.10) for three different values of the selection coefficient: $s = 0.2$, $s = 0.4$, and $s = 0.6$.

dominant allele is lethal, such that anyone having either the AA or Aa genotype will be eliminated in the first generation, and only those with the recessive homozygote aa will survive. We can also see this effect mathematically by using complete selection ($s = 1$) in equation (6.10), which shows that if we start with allele frequency p, the frequency of the dominant allele in the next generation will be

$$p' = \frac{p(1 - s)}{1 - s + sq^2} = \frac{p(1 - 1)}{1 - 1 + (1)q^2} = 0$$

This is the only way that selection can remove an allele in a single generation.

C. Selection with Codominant Alleles

The examples presented thus far have involved a dominant allele and a recessive allele. The next model considers that happens when the two alleles are codominant and there is selection against one of the alleles. Under codominance, the heterozygote Aa will show the effect of both alleles. One way to model selection in this type of situation is consider selection for the A allele and selection against the a allele. This is done by assigning the highest fitness to individuals with two A alleles (genotype AA), the lowest fitness to someone with no A alleles (genotype aa), and an intermediate fitness to those with one A allele (genotype Aa). The fitness values in this case are

$$w_{AA} = 1$$

$$w_{Aa} = 1 - \frac{s}{2}$$

$$w_{aa} = 1 - s$$

Note that the fitness of the heterozygote is the average of the fitness values of the two homozygotes, thus meeting our criterion that the fitness of the heterozygote is intermediate:

$$\frac{w_{AA} + w_{aa}}{2} = \frac{(1) + (1-s)}{2} = \frac{2-s}{2} = 1 - \frac{s}{2}$$

Mean fitness is derived by substituting these values into equation (6.1), giving

$$\overline{w} = 1 - sq \tag{6.12}$$

(see Appendix 6.4). This equation shows us that for any value of s, mean fitness will increase as q decreases. Selection against the a allele will lead to its decline even under codominance.

The frequency of the a allele in the next generation is specifically derived by substituting the fitness values into equation (6.3), which gives

$$q' = \frac{q - sq(1+q)/2}{1 - sq} \tag{6.13}$$

The amount of change in the a allele per generation is now obtained from equation (6.5), giving

$$\Delta q = \frac{-spq/2}{1 - sq} \tag{6.14}$$

(see Appendix 6.4). This equation has a negative sign in the numerator, which means that the frequency of the a allele decreases with each generation. This decrease will continue until $\Delta q = 0$, which will occur when $q = 0$. As with selection against the recessive homozygote, this is a mathematical limit.

Figure 6.6 shows selection against the codominant allele a for three different values of the selection coefficient. As with selection against the recessive homozygote (Figure 6.4), the allele frequency decreases rapidly at first, approaching a value of zero. Note, however, that the rate of change of the examples shown in Figure 6.6 is faster (i.e., descent of the curves is steeper) than those shown in Figure 6.4 for the same selection coefficients. For example, under selection against the recessive homozygote for $s = 0.2$, the allele frequency drops from $q = 0.5$ to about $q = 0.2$ (Figure 6.4) after 20 generations, whereas the frequency drops to $q = 0.1$ after 20 generations of selection against the codominant allele (Figure 6.6). The reason for this is that the fitness of the heterozygote is lower for the case of selection against the codominant allele. Under selection against the recessive homozygote, no heterozygotes are selected against (because fitness $= 1$), whereas some heterozygotes are selected against under codominance (where fitness $= 1 - s/2$).

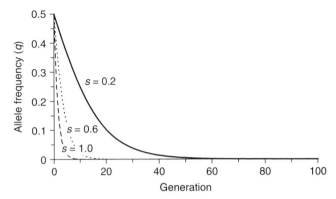

FIGURE 6.6 Selection against the codominant allele *a* for different values of the selection coefficient. Initial allele frequencies are $p = 0.5$ and $q = 0.5$ in each case. Fitness values are $w_{AA} = 1$, $w_{Aa} = 1 - (s/2)$, and $w_{aa} = 1 - s$. Allele frequencies were derived using equation (6.13) for three different values of the selection coefficient: $s = 0.2$, $s = 0.6$, and $s = 1.0$. Compare the rates of change here with the case of selection against the recessive homozygote shown in Figure 6.4.

D. Selection against the Heterozygote

The examples presented thus far have shown selection against one of the homozygotes relative to other genotypes. The results have been intuitive; selection against an allele causes it to decrease over time (and, from the perspective of the other allele, selection for an allele causes it to increase over time). The case of selection against the heterozygote is a bit different. In order to model this type of selection, we assign the heterozygote a lower fitness than the two homozygotes. The simplest model of this type uses fitness values of

$$w_{AA} = 1$$
$$w_{Aa} = 1 - s$$
$$w_{aa} = 1$$

The mean fitness is determined by substituting these values into equation (6.1), giving

$$\overline{w} = 1 - 2spq \tag{6.15}$$

(see Appendix 6.5 for derivation). The relationship between mean fitness and allele frequency is more complicated than in the previous examples of natural selection, where fitness increased (or decreased) with allele frequency. Here, the relationship is parabolic, which is easily seen when substituting $(p = 1 - q)$ into equation (6.15), giving

$$\overline{w} = 1 - 2sq + 2sq^2$$

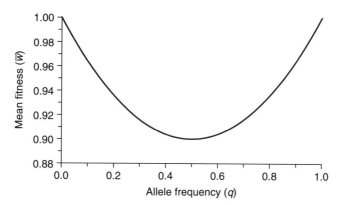

FIGURE 6.7 Mean fitness as a function of allele frequency under selection against the heterozygote. Mean fitness was derived using equation (6.15) with a selection coefficient of $s = 0.2$. Note that mean fitness reaches a maximum for two values of allele frequency, $q = 0$ and $q = 1$, and a minimum for $q = 0.5$.

The parabolic relationship is easily seen in Figure 6.7, which plots mean fitness as a function of q for a given selection coefficient. It is clear that mean fitness is at a maximum for two different allele frequencies: $q = 0$ and $q = 1$. Note also that the minimum fitness is found at an allele frequency of $q = 0.5$.

The frequency of the a allele in the next generation is obtained using equation (6.3), giving

$$q' = \frac{q - spq}{1 - 2spq} \tag{6.16}$$

The amount of change in the a allele per generation is now obtained from equation (6.5), giving

$$\Delta q = \frac{spq(q - p)}{1 - 2spq} \tag{6.17}$$

(see Appendix 6.5 for derivations). Note that the sign (positive or negative) of this equation is not clear, but depends on the exact value of the term $(q - p)$, which, in turn, depends on whether q is greater or less than p. If q is greater than p, then this term is positive, and the frequency of q will increase in the next generation. On the other hand, if q is less than p, then the term $(q - p)$ is negative, and the frequency of q will decrease in the next generation. Thus, the frequency of an allele could increase or decrease, depending on the initial allele frequency!

To see this, let us try an example. Let the selection coefficient be $s = 0.2$ and let $q = 0.6$ (which means that $p = 0.4$). Plugging these values into equation (6.17) gives

$$\Delta q = \frac{(0.2)(0.4)(0.6)(0.6 - 0.4)}{1 - 2(0.2)(0.4)(0.6)} = \frac{0.0096}{0.904} = 0.0106$$

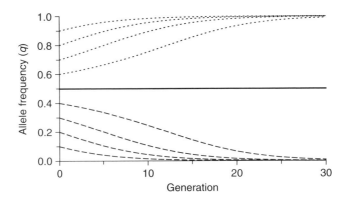

FIGURE 6.8 Selection against the heterozygote for different staring values of q. Fitness values are $w_{AA} = 1$, $w_{Aa} = 1 - s$, and $w_{aa} = 1$. Allele frequencies were derived using equation (6.16) for $s = 0.2$ and nine different initial values of q: 0.1, 0.2, 0.3, 0.4, 0.5, 0.6, 0.7, 0.8, and 0.9. Dotted lines show selection when the initial allele frequency is greater than 0.5, and dashed lines show selection when the initial allele frequency is less than 0.5. The solid line shows the case where the initial allele frequency is exactly $q = 0.5$.

which is a positive value, meaning that the frequency of q increases to $0.6 + 0.0106 = 0.6106$. Now, let us look at the case where s is still equal to 0.2, but now $q = 0.4$ and $p = 0.6$. Here, equation (6.17) gives

$$\Delta q = \frac{(0.2)(0.6)(0.4)(0.4 - 0.6)}{1 - 2(0.2)(0.6)(0.4)} = \frac{-0.0096}{0.904} = -0.0106$$

which has a negative value, meaning that the frequency of q decreases from 0.4 to $0.4 - 0.0106 = 0.3894$.

    The effect of initial allele frequency is shown in Figure 6.8 for a number of different starting values of q. This graph shows that when the initial allele frequency is greater than 0.5, the effect of selection is to *increase* the allele frequency until a value of $q = 1.0$ is reached. The graph also shows that when the initial allele frequency is less than 0.5, selection acts to *decrease* the frequency allele until a value of $q = 0.0$ is reached. The one exception is when the initial allele frequency is *exactly* 0.5, in which case there will be no change!

    Equation (6.17) shows how the equilibrium value of the alleles also depends on the initial allele frequency. We determine the equilibrium values by setting equation (6.17) equal to zero, which will be true when $q = 0$ or when $p = 0$ (which means that $q = 1$). If q is less than 0.5, then Δq will be negative and the allele frequency will decrease until $q = 0$. If, however, q is greater than 0.5, then Δq will be positive and increase until $q = 1$. There is also a third equilibrium value. Note that equation (6.17) will be equal to zero when the term $(q - p)$ is equal to zero, which means that $p = q = 0.5$. As we have seen in Figure 6.8, when the initial allele frequency is equal to 0.5, there will be no further change. It would be very rare, however, for such a condition to persist for long, as mutation, drift, and/or gene flow would soon tip the allele frequency to slightly more (or less) than 0.5, and the selection would continue to move the allele frequency toward an equilibrium of $q = 0$ or $q = 1$.

An intuitive way of seeing the effect of selection against the heterozygote is to consider the following analogy. Assume that you are in a class with 40 students. There is a table in the front of the room that contains 60 apples and 40 oranges. Now, imagine that each student comes to the front of the class and takes one apple and one orange (the two types of fruit chosen are analogous to a heterozygote with two different alleles). After the first student takes an apple and an orange, there will be 59 apples and 39 oranges remaining. After the second student takes an apple and an orange, there will be 58 apples and 38 oranges. This will continue until the last of the 40 students has taken an apple and an orange. There are now 20 apples and no oranges left on the table. In other words, we ran out of oranges first because there were fewer oranges to start with. If we repeated this thought experiment, but this time imagined 40 apples and 60 oranges, we would run out of apples first because there are fewer apples to start with. The same process occurs with selection against the heterozygote. Every time an individual with the heterozygote does not survive and reproduce, *both* an A allele and an a allele are removed from the population. The eventual fate depends on how many A and a alleles there were to begin with. If there are more A alleles, then selection will lead to only A alleles. If there are more a alleles, then selection will lead to only a alleles.

E.   Selection for the Heterozygote

All of the previous examples have involved selection that has led to an increase in allele frequency toward an equilibrium value of 1.0, or a decrease in allele frequency toward an equilibrium value of 0.0. Ultimately, an allele either is eliminated from the population or reaches fixation within the population. There are no intermediate equilibrium values (other than the unrealistic case of $p = q = 0.5$ under selection against the heterozygote). Selection *for* the heterozygote is different, because it leads to an equilibrium value that lies somewhere between 0.0 and 1.0. Instead of one allele being favored completely over the other, a *balance* in allele frequencies is reached. In fact, selection for the heterozygote is often referred to as **balancing selection**.

In this model, the heterozygote has the highest fitness and the homozygotes are assigned lower fitness values. In order to show how different balances occur, we assign different selection coefficients for the two homozygotes. Here, s is the selection coefficient of the homozygous genotype AA and we use the letter t for the selection coefficient of the homozygous genotype aa. The fitness values of the three genotypes are

$$w_{AA} = 1 - s$$
$$w_{Aa} = 1$$
$$w_{aa} = 1 - t$$

The mean fitness is determined by substituting these values into equation (6.1), giving

$$\overline{w} = 1 - sp^2 - tq^2 \qquad (6.18)$$

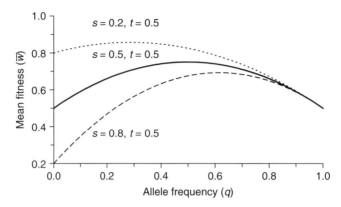

FIGURE 6.9 Mean fitness as a function of allele frequency under selection for the heterozygote. Mean fitness was derived using equation (6.18) for three sets of selection coefficients s and t. In each case, the relationship between fitness and allele frequency is parabolic, and fitness is at a maximum for an intermediate allele frequency and not for the extreme values of $q = 0$ or $q = 1$.

(see Appendix 6.6). The relationship between fitness and allele frequencies is parabolic, but the exact shape depends on the values of s and t. Figure 6.9 provides an example of this relationship for three different sets of values of s and t. Each curve is parabolic, but each reaches a maximum fitness at a different value of q. Note that, unlike previous examples of selection, the maximum fitness values are *not* found at the extreme values of $q = 0$ or $q = 1$.

The frequency of the a allele in the next generation is derived from equation (6.3) to give

$$q' = \frac{q - tq^2}{1 - sp^2 - tq^2} \tag{6.19}$$

and the amount of allele frequency change per generation is obtained from equation (6.5), giving

$$\Delta q = \frac{pq[sp - tq]}{1 - sp^2 - tq^2} \tag{6.20}$$

(see Appendix 6.6 for derivations). Note that the sign of the numerator could be positive or negative, indicating that in some cases the allele frequency will increase, and in some cases it will decrease. If $sp > tq$, then the numerator is positive and the allele frequency will increase. If $sp < tq$, then the numerator is negative and the allele frequency will decrease. Equilibrium occurs when $\Delta q = 0$, which will happen when $sp = tq$. This relationship facilitates calculation of the allele frequencies at equilibrium by substituting $(1 - q)$ for p, which gives

$$sp = tq$$

$$s(1 - q) = tq$$

Solving for q gives

$$q = \frac{s}{s+t} \tag{6.21}$$

Because $p = 1 - q$, this means that the equilibrium frequency of p will be

$$p = \frac{t}{s+t} \tag{6.22}$$

(Readers who know calculus may be interested in a calculus-based derivation of the equilibrium allele frequency in Appendix 6.7.)

Let us consider an example where $s = 0.2, t = 0.5$, and the initial allele frequency is $q = 0.1$. In this case, the fitness of genotype AA is $(1 - 0.2) = 0.8$, which is lower than the fitness of the heterozygote ($= 1$, by definition), but higher than the fitness of genotype aa, which is $(1 - 0.5) = 0.5$. Because the fitness of genotype aa is the lowest, we might expect the frequency of the a allele to decrease because a alleles are being removed from the population. However, because the heterozygote has the highest fitness, a alleles are also being put back into the population. Further, some A alleles are being removed by selection against genotype AA, but are also being maintained in the population because of the higher fitness of the heterozygote. In essence, both alleles are being selected for and against in this case. You can see why this is called *balancing selection*!

If we use the hypothetical values of $s = 0.2, t = 0.5$, and an initial value of $q = 0.1$ with equation (6.20), we see that the critical term $[sp - tq]$ is equal to

$$[(0.2)(1 - 0.1) - (0.5)(0.1)] = 0.13$$

which is positive. This means that q will increase over time, and will reach the equilibrium value from equation (6.21) of

$$q = \frac{s}{s+t} = \frac{0.2}{0.2 + 0.5} = 0.286$$

Now, let's use the same selection coefficients $s = 0.2$ and $t = 0.5$, but start instead with an initial allele frequency of $q = 0.9$. The critical term $[sp - tq]$ is now equal to

$$[(0.2)(1 - 0.9) - (0.5)(0.9)] = -0.43$$

which is negative, which means that the allele frequency will decrease. However, the equilibrium value from equation (6.21) is still the same. Thus, selection will decrease the allele frequency until the equilibrium of $q = 0.286$ is reached. Figure 6.10 shows both of these examples, the one starting with $q = 0.1$ and the one starting with $q = 0.9$. Both converge to the same equilibrium. Thus, the direction of allele frequency change depends on the initial allele frequencies, but the equilibrium is the same and is determined by the selection coefficients.

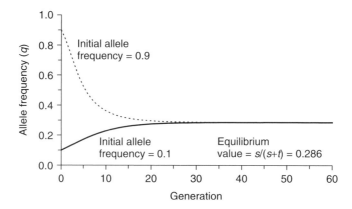

FIGURE 6.10 Selection for the heterozygote for two different starting values of q. Fitness values are $w_{AA} = 1 - s, w_{Aa} = 1$, and $w_{aa} = 1 - t$. Allele frequencies were derived using equation (6.19) for $s = 0.2$ and $t = 0.5$. The solid line shows the pattern of allele frequency change starting with an initial allele frequency of $q = 0.1$, and the dotted line shows the pattern of allele frequency change starting with an initial allele frequency of $q = 0.9$.

IV. OTHER ASPECTS OF SELECTION

The remainder of this chapter extends the basic models of selection to deal with other aspects of natural selection, including interaction with other evolutionary forces (mutation, drift), interaction with inbreeding, and implications of selection for the evolution of quantitative traits.

A. Selection and Mutation

Now that we have covered two evolutionary forces in detail, mutation and selection, it is time to consider how they can interact. If we focus only on these two forces at the moment, ignoring genetic drift and gene flow, we can picture microevolution as a process where new mutations introduce genetic variation into a population in small amounts and where natural selection determines the fate of these new mutants.

Selection for a Mutation

Intuitively, a helpful mutant allele will increase in frequency. An example is shown in Figure 6.11. This particular simulation starts with allele frequencies of $p = 1.0$ and $q = 0.0$. Each generation, some A alleles mutate into a alleles at a mutation rate of $\mu = 0.00001$. Genotype fitness values were set to show what happens when the alleles are codominant where the mutant allele is favored. In this example, those with two mutant alleles have the highest fitness ($w_{aa} = 1.0$), those with no mutant alleles have the lowest fitness ($w_{AA} = 0.8$), and those with one mutant allele have an intermediate fitness ($w_{Aa} = 0.9$). The frequency of the mutant allele stays low for a number of generations because the proportion of individuals with favored genotypes is low. As the number of individuals with genotype Aa or aa increases, the chance for selection increases, leading to an exponential increase after about

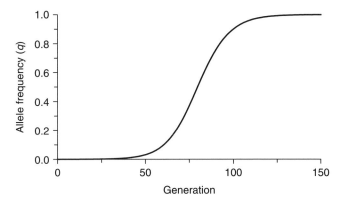

FIGURE 6.11 Selection for an advantageous mutant allele. Initial allele frequencies are $p = 1$ and $q = 0$. With each generation there is mutation from the A allele to the a allele with mutation rate $\mu = 0.00001$. Selection favors the mutant allele; those with two mutant alleles (aa) have the highest fitness ($= 1$), those with one mutant allele (Aa) have the next highest fitness ($= 0.9$), and those with no mutant alleles (AA) have the lowest fitness ($= 0.8$).

60 generations. Selection is particularly rapid after that, and the frequency of the mutant allele is over 90% after 100 generations. The rate of selection would be even higher if the mutant allele were dominant.

Selection against a Mutation

It might seem reasonable to assume that because allele frequency decreases when there is selection against that allele, selection against a mutation will totally remove the mutation in a population. Actually, this does not happen. Instead, an equilibrium is reached between selection against the mutation (which removes the allele) and mutation (which adds the allele). An example is shown in Figure 6.12 for mutation to a harmful recessive allele (A to a) with a mutation rate of $\mu = 0.0001$ and a selection coefficient of $s = 0.2$ for the recessive homozygote. The frequency of the mutation allele increases with time at first and then levels off to a value of roughly $q = 0.022$. This increase would occur even if there were complete selection against the recessive homozygote, as the heterozygotes would carry the mutant allele without harm.

Why does this happen? When q is very low, there are few individuals with the genotype aa to be selected against. Over time, the mutations accumulate in the population due to continued mutation of A alleles into a alleles, until a point is reached where there are enough individuals with genotype aa to be selected against, which offsets the continued mutation, and an equilibrium is reached. As an analogy, consider the example of the paper cup from Chapter 5. When you place a paper cup with a small hole in the bottom under a water faucet set to a slow drip, the water entering the cup will cause the water level to rise, whereas the hole in the bottom of the cup will cause the water level to drop. If you do this just right, you will be able to find the correct amount of water that allows a balance between the water entering the cup and the water leaving the cup. The water entering the cup is analogous to mutation, and the water leaving the cup through

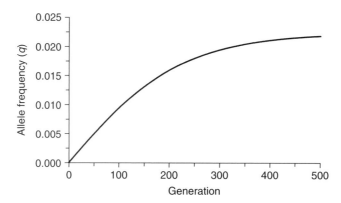

FIGURE 6.12 Balance between mutation to a recessive allele and selection against the recessive homozygote. Mutation rate (for $A \rightarrow a$) is $\mu = 0.0001$, and the selection coefficient for the recessive homozygote is $s = 0.2$.

the hole at the bottom is analogous to selection against the mutation. The water level at equilibrium (where the water leaving balances the water entering the cup) is analogous to the equilibrium allele frequency where mutation and selection are balanced.

In the case of selection against the recessive homozygote, this balance is approximately equal to $q = \sqrt{\mu/s}$, where μ is the mutation rate from A to a and s is the selection coefficient for the recessive homozygote (see Appendix 6.8 for derivation of this approximation). In Figure 6.12, the mutation rate is $\mu = 0.0001$ and the selection coefficient is $s = 0.2$, giving an approximate equilibrium value of $q = \sqrt{\mu/s} = \sqrt{\frac{0.0001}{0.2}} = \sqrt{0.0005} = 0.0224$. The approximation formula shows us that the lower the mutation rate and/or the higher the selection coefficient, the lower the equilibrium frequency. For example, with a mutation rate of $\mu = 0.000001$ and a selection coefficient of $s = 0.5$, the equilibrium frequency would be approximately 0.0014. Although selection will keep the frequency of the mutation low, it is *not* zero. Even when the mutation is lethal ($s = 1$), the equilibrium frequency will not be zero, but instead will be $\sqrt{\mu}$. Heterozygotes will continue to carry the recessive lethal allele, allowing the frequency to build up to this low, but nonzero, equilibrium.

What about a dominant mutation? An example is shown in Figure 6.13 for a mutation to a harmful dominant allele (a into A) with a mutation rate of 0.0001 and a selection coefficient of $s = 0.2$ for the recessive homozygote. The frequency of the mutant allele increases rapidly at first, but then levels off to an equilibrium value after about 20 generations. The equilibrium is approximately $p \approx \mu/s$ (see Appendix 6.9 for derivation). This equilibrium frequency is smaller than for the previous case of mutation and selection against the recessive allele, which makes sense as there are more opportunities for selection against a dominant allele (both genotypes AA and Aa are affected). For Figure 6.13, the approximate equilibrium frequency is $p \approx \mu/s = \frac{0.0001}{0.2} = 0.0005$.

The equilibrium formula also shows that even if there is total selection against the dominant allele ($s = 1$ for both genotype AA and genotype aa), the frequency

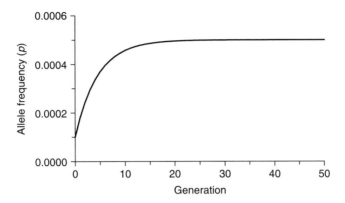

FIGURE 6.13 Balance between mutation to a dominant allele and selection against the dominant allele. Mutation rate (for $a \rightarrow A$) is $\mu = 0.0001$, and the selection coefficient for the genotypes with the dominant allele (AA and Aa) is $s = 0.2$.

of the mutant allele is still $p = \mu$. Thus, even though all individuals in any given generation with a lethal dominant allele are selected out, the presence of new mutations in any generation keeps the allele frequency above zero.

B. Selection and Genetic Drift

How do selection and drift interact? Which of these forces has the major effect on allele frequency change, and under what conditions? Of particular interest is the case where selection and drift act in opposition to each other. Can the frequency of an allele that has been selected against increase because of genetic drift? The obvious answer to this question is that the fate of an allele will depend on the relative strength of selection and drift.

Computer simulation gives us an idea of different ways in which selection and drift can interact. Here, I conducted three simulations, each using a small population size of $N = 20$ but varying the fitness values of the genotypes. This small population size means that there is a lot of potential for genetic drift, and by varying the fitness values, we can get an idea of the ways in which drift and selection interact. Each simulation was run 1000 times, all starting with an initial frequency of the a allele of $q = 0.5$, and allowed to run for 100 generations. The distribution of allele frequencies after 100 generations is shown in Figure 6.14 for the three simulation experiments.

The first simulation of drift and selection used very minor differences in fitness values: $w_{AA} = 1.0, w_{Aa} = 0.995$, and $w_{aa} = 0.9$. Even with the very small differences in fitness values, this type of selection slowly decreases the frequency of the a allele over time. For example, if there were no drift, the frequency of the a allele would be $q = 0.377$ after 100 generations. However, Figure 6.14a shows a distribution characteristic of genetic drift with a high proportion of final values of q that have reached extinction or fixation. Here, selection has practically no impact relative to genetic drift, which is expected given the very small differences in fitness. Figure 6.14b shows the opposite pattern. This simulation used large differences

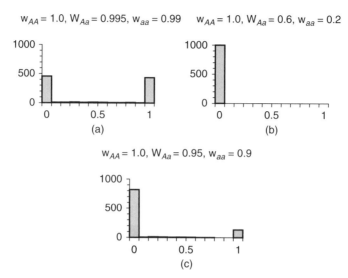

$w_{AA} = 1.0, W_{Aa} = 0.995, w_{aa} = 0.99$ $w_{AA} = 1.0, W_{Aa} = 0.6, w_{aa} = 0.2$

(a)   (b)

$w_{AA} = 1.0, W_{Aa} = 0.95, w_{aa} = 0.9$

(c)

FIGURE 6.14   Distributions of allele frequencies under the simulation of selection and drift. Each graph shows the distributions of the *a* allele after 100 generations from 1000 runs of natural selection and genetic drift in a population of $N = 20$ and an initial allele frequency of $q = 0.5$. The three distributions use different fitness values: (a) $w_{AA} = 1.0, w_{Aa} = 0.995, w_{aa} = 0.9$; (b) $w_{AA} = 1.0, w_{Aa} = 0.6, w_{aa} = 0.2$; (c)$w_{AA} = 1.0, w_{Aa} = 0.95, w_{aa} = 0.9$. In Figure 6.14a, drift has dominated the final distribution of allele frequencies, whereas in Figure 6.14b, selection has led to the extinction of all *a* alleles. In Figure 6.14c, both selection and drift have had an impact. Under drift, some runs led to the fixation of the *a* allele, but not as many as in Figure 6.14a, where drift overrode the effects of selection.

in fitness values: $w_{AA} = 1.0, w_{Aa} = 0.6$, and $w_{aa} = 0.2$. Here, all of the 1000 runs wound up with a value of $q = 0$ by 100 generations. In this simulation, selection clearly dominated allele frequency change and drift had no impact. Figure 6.14c shows an intermediate case. The fitness values for the third simulation used larger differences in fitness values than in the first simulation, but nowhere near as large as in the second simulation: $w_{AA} = 1.0, w_{Aa} = 0.95$, and $w_{aa} = 0.9$. The graph shows that a large majority of the runs led to the loss of the *a* allele over time, as expected (under selection alone, the expected frequency of the *a* allele after 100 generations is $q = 0.0057$). Some runs wound up with higher values of q, and a substantial minority of the runs showed fixation of the *a* allele by 100 generations! For some values of fitness and population size, it is possible for a disadvantageous allele (one being selected against) to not only increase in size but also become fixed through genetic drift.

The interaction between natural selection and genetic drift can be extended further by also considering mutation. As was shown in Chapter 5, the neutral theory of evolution shows that although many new mutations are lost because of genetic drift, some new mutants will reach fixation. What happens if there is selection *against* a new mutant? Can drift overcome the effects of selection when the frequency of the new mutation is so close to zero to begin with? Geneticist Tomoko Ohta had developed the **nearly neutral theory of evolution** that examines this possibility (Ohta 1992). If selection is weak (but not absent) and the population size is small, then drift can have a major impact on allele frequency, causing results that are similar to alleles that are completely neutral. The population size is a critical parameter in the nearly neutral theory; weak selection can dominate in large populations, whereas in small populations drift will have a greater impact.

C. Selection and Inbreeding

Another type of interaction is natural selection under inbreeding. We know from Chapter 3 that inbreeding does not directly change allele frequencies, but can have an impact on the effect of the evolutionary forces. In the case of selection, inbreeding can affect the rate of change in allele frequencies because inbreeding results in proportionately more homozygotes and fewer heterozygotes. The general model of natural selection developed in this chapter assumed that genotype frequencies before selection were at Hardy–Weinberg equilibrium. We can relax this assumption by using the genotype frequencies expected under inbreeding from equation (3.4) in Chapter 3. Figure 6.15 illustrates the effect of inbreeding on selection against the recessive homozygote by comparing the change in allele frequency over time under three different levels of inbreeding: $F = 0$, where there is no inbreeding; $F = 0.015625$, where everyone in the population mates with a second cousin; and $F = 0.0625$, where everyone in the population mates with a first cousin. It is clear from Figure 6.15 that the higher the level of average inbreeding, then the quicker the reduction in the frequency of the recessive allele. For example, starting from an allele frequency of $q = 0.1$, it takes 20 generations under no inbreeding to reach a value of $q = 0.05$, but only 11 generations for $F = 0.0625$. Nonetheless, given the relatively low levels of average inbreeding

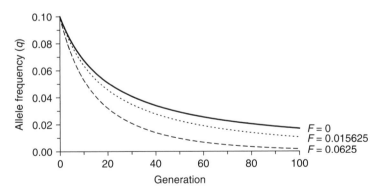

FIGURE 6.15   Selection against the recessive homozygote under different levels of inbreeding. The initial allele frequency of the recessive allele is $q = 0.1$. Population inbreeding levels of $F = 0.015625$ and $F = 0.0625$ are compared with no inbreeding ($F = 0$). These levels correspond to the case where everyone in the population mates with their second cousin ($F = 0.015625$) and the case where everyone mates with their first cousin ($F = 0.0625$). Inbreeding increases the rate of selection against the recessive homozygote; at any given generation, the higher the value of F, the lower the allele frequency.

in most human populations (see Chapter 3), the actual impact of inbreeding on selection is likely low.

D. Natural Selection and Quantitative Traits

Although the discussions so far have focused on simple models using a single locus with two alleles, a number of extensions can be made for complex models, including selection on quantitative traits, such as height and skin color. A detailed account of the modeling of natural selection on quantitative genetics is beyond the scope of this book, and interested readers are referred to short reviews by Konigsberg (2000) and Mielke et al. (2011), and the classic, more specialized text by Falconer and Mackay (1996). It is useful, however, to consider in a broad sense the general types of selection on quantitative traits relevant to humans.

One type of selection is **stabilizing selection**, where those with more extreme phenotypes are selected against, and selection is instead for those with intermediate phenotypes. Because selection is against *both* extremes, the average phenotype is maintained (stabilized) over time. An example of this type of selection in human populations is for birth weight. Studies of birth weight have shown that babies that are have either very low or very high birth weights have a lower probability of surviving than do those with average birth weight (e.g., Karn and Penrose 1952). The other major type of selection of interest in human populations is **directional selection**, where one extreme of the range of phenotypes is selected for, producing a shift in average value. On example from human evolution is the noticeable increase in brain size over the past 2 million years. Another example, illustrated in the next chapter, is the change to lighter skin color among humans whose ancestors moved into northern latitudes.

E. Natural Selection: Theory and Reality

Models of natural selection are elegant descriptions to use in what-if scenarios regarding the impact of mode of selection and variation in fitness rates. Such models are very useful in providing a baseline understanding of how selection could work under different sets of conditions. What about the real world? How well do these models apply to actual case studies of selection, particularly in human populations? Application to actual studies of selection in human populations is the focus of the next chapter, but we need to raise a few questions and caveats at this point before moving on to these case studies.

The case studies discussed in this chapter provide powerful examples of how selection can produce major changes in the gene pool over time. Even small differences in fitness can add up to major differences over a relatively small number of generations (small by evolutionary standards—a few dozen generations is a short period of time in evolution, although quite long in terms of our own lifetime). In an idealized world, mutations occur with each generation and their fate is ultimately determined by selection for or against the new mutants. From the perspective of the average human lifespan, such changes occur at a glacial pace, but quickly add up to produce significant change.

Although this model is very useful, we have to keep in mind that the actual situation with regard to selection might not always be that simple. For one thing, the mathematical models that we have used in this chapter assume constancy of fitness values over dozens of generations. In reality, fitness values can often change over time, particularly when the environment changes. This does not invalidate the use of simple mathematical models for understanding selection, but instead means that we need to consider violation of basic assumptions in the real world. Much of science consists in developing a simple model, testing its fit in the real world, and then explaining why and how it does or does not fit.

The case of the peppered moth described in earlier chapters is an excellent example. The difference between light-colored and dark-colored moths is due to a single locus with two alleles—light and dark—where the light allele is recessive and the dark allele is dominant. Initially, selection was against the dark-colored moths because they stood out on the light-colored lichen of the trees, making them more likely to be eaten by birds. This is selection against a dominant allele, which leads ultimately to a balance between mutation and selection, and a very low proportion of dark moths. Environmental conditions then changed; industrial pollution led to lichen dying and exposing the dark tree trunks. Now, light-colored moths stood out and were more likely to be eaten, and dark-colored moths had the advantage. Therefore, the dark mutant allele was no longer selected against. Selection *for* the dominant allele took place, leading quickly to the virtual replacement by dark alleles.

The point here is that the environment changed, and so did the fitness values. What might be harmful in one environment could be helpful in another environment. If the environment changes, then so might fitness values, leading to a different trajectory of evolution through natural selection. As will be shown in examples in the next chapter, it is critical to keep in mind the possibility of changes in fitness values in human populations because our rapid cultural change alters how we interact with the environment very rapidly. This is very apparent when considering some allele frequencies that have changed considerably in some human populations even during the last 10,000 years (again, this seems like a very long time to you and me, but it is a mere geologic instant). Humans have the capability for rapid cultural and genetic change.

There is continuing debate over the relative impact of selection in terms of evolutionary change. For example, Weiss and Buchanan (2009) argue that most selection consists of weeding out harmful mutations rather than generating new genetic combinations. In this sense, they suggest that it would make more sense to describe natural selection as a process of "failure of the frail" rather than "survival of the fittest."

V. SUMMARY

This chapter has focused on the underlying mathematics of natural selection by using a simple simulation of selection to develop a general model of allele frequency change under natural selection. This general model was used with different fitness values to understand the general principles of several basic models of natural selection, including selection against the recessive homozygote, selection against a dominant allele, selection against a codominant allele, selection against the heterozygote, and selection for the heterozygote (balancing selection). Apart from balancing selection, the other forms of selection lead to the elimination of one allele over time, although this can take many generations (apart from selection against a lethal dominant allele). Balancing selection, on the other hand, will reach an equilibrium allele frequency that is not 0 or 1.

Mutation can interact with selection in one of two ways. An advantageous allele will be selected for, increasing in frequency to reach either fixation or, in the case of balancing selection, an intermediate equilibrium value. Selection against a mutant allele will reduce the frequency of that allele, but not to zero, because new mutations are likely to be introduced into the population with each generation. An equilibrium between mutation and selection is expected here, leading to a very low, but nonzero, allele frequency. Natural selection also interacts with selection, with the relative impact of selection and drift depending on the intensity of selection and the population size. In small populations with minor differences in fitness between genotypes, the frequency of a disadvantageous allele can increase and even reach fixation as a result of genetic drift. Inbreeding interacts with selection by changing the speed of selection. For example, when selection is against the recessive homozygote, the effect of selection increases with each generation because of the increased number of homozygotes under inbreeding. Consequently, the frequency of the recessive allele decreases over time at a faster rate than expected under random mating.

This chapter has been theoretical by design. Although the models used in this chapter are often simplistic, they do provide us with some general ideas of how selection can operate. The next chapter moves from the theoretical to the real by providing some selected examples of how natural selection is measured and analyzed in human populations.

APPENDIX 6.1 DERIVATION OF THE AMOUNT OF CHANGE IN ALLELE FREQUENCIES PER GENERATION (Δp AND Δq) FOR A GENERAL MODEL OF NATURAL SELECTION

This derivation follows the presentation outlined by Ayala (1982), which I find particularly clear. We will start with the case for p, the frequency of the A allele. After one generation of natural selection, the mean fitness is, from equation (6.1)

$$\overline{w} = p^2 w_{AA} + 2pq w_{Aa} + q^2 w_{aa}$$

and the allele frequency is, from equation (6.2)

$$p' = \frac{p(pw_{AA} + qw_{Aa})}{\overline{w}}$$

The change in allele frequency is

$$\Delta p = p' - p$$

which, after putting these two equations together, gives

$$\Delta p = \frac{p(pw_{AA} + qw_{Aa})}{\overline{w}} - p$$

The second part of this equation (p) is multiplied by $\overline{w}/\overline{w}$ to get a common denominator, giving

$$\Delta p = \frac{p(pw_{AA} + qw_{Aa}) - p\overline{w}}{\overline{w}}$$

We substitute equation (6.1) for the mean fitness in the numerator, giving

$$\Delta p = \frac{p(pw_{AA} + qw_{Aa}) - p(p^2w_{AA} + 2pqw_{Aa} + q^2w_{aa})}{\overline{w}}$$

After factoring p, this gives

$$\Delta p = \frac{p[pw_{AA} + qw_{Aa} - p^2w_{AA} - 2pqw_{Aa} - q^2w_{aa}]}{\overline{w}}$$

This equation is rearranged by collecting terms for the same fitness values (all terms for w_{AA}, all terms for w_{Aa}, and all terms for w_{aa}), which gives

$$\Delta p = \frac{p[p(1 - p)w_{AA} + q(1 - 2p)w_{Aa} - q^2w_{aa}]}{\overline{w}}$$

Note that because $p + q = 1$, the term $(1 - 2p)$ can be rewritten as $(p + q - 2p) = (q - p)$. This relationship, along with the relationship $(1 - p = q)$, allows us to rewrite the equation as

$$\Delta p = \frac{p[pqw_{AA} + qw_{Aa}(q - p) - q^2w_{aa}]}{\overline{w}}$$

Factoring out q gives

$$\Delta p = \frac{pq[pw_{AA} + w_{Aa}(q - p) - qw_{aa}]}{\overline{w}}$$

which, when multiplied out, gives

$$\Delta p = \frac{pq[pw_{AA} + qw_{Aa} - pw_{Aa} - qw_{aa}]}{\overline{w}}$$

The final step is to collect terms for p and q, giving

$$\Delta p = \frac{pq[p(w_{AA} - w_{Aa}) + q(w_{Aa} - w_{aa})]}{\overline{w}}$$

which is equation (6.4) in the main (chapter) text. The same process can be used to describe the change in $q(\Delta q = q' - q)$ from one generation of natural selection, which will give

$$\Delta q = \frac{pq[p(w_{Aa} - w_{AA}) + q(w_{aa} - w_{Aa}]}{\overline{w}}$$

which is equation (6.5) in the text.

APPENDIX 6.2 DERIVATION OF FORMULAS FOR SELECTION AGAINST THE RECESSIVE HOMOZYGOTE

This section provides the derivation of equations (6.6)–(6.8).

In this model, the A allele is dominant and the a allele is recessive. Under selection against the recessive homozygote, the fitness values for the three genotypes are $w_{AA} = 1, w_{Aa} = 1$, and $w_{aa} = 1 - s$. The mean fitness for a general model of natural selection, from equation (6.1)

$$\overline{w} = p^2 w_{AA} + 2pq w_{Aa} + q^2 w_{aa}$$

Substituting the fitness values for selection against the recessive homozygote gives

$$\overline{w} = p^2(1) + 2pq(1) + q^2(1 - s)$$

When multiplied out, this gives

$$\overline{w} = p^2 + 2pq + q^2 - sq^2$$

Because the first three terms add up to 1 ($p^2 + 2pq + q^2 = 1$), the mean fitness is

$$\overline{w} = 1 - sq^2$$

which is equation (6.6) in the main text.

We now derive the formula for the allele frequency in the next generation (q') by substituting the fitness values into equation (6.3), which gives

$$q' = \frac{q(pw_{Aa} + qw_{aa})}{\overline{w}}$$

$$= \frac{q[p(1) + q(1 - s)]}{\overline{w}}$$

Multiplying this out, and substituting the value for mean fitness from above, gives

$$q' = \frac{q[p + q - sq]}{1 - sq^2}$$

Because by definition $(p + q) = 1$, this equation reduces to

$$q' = \frac{q[1 - sq]}{1 - sq^2} = \frac{q - sq^2}{1 - sq^2}$$

which is equation (6.7) in the text.

The amount of allele frequency change per generation ($\Delta q = q' - q$) is now derived using equation (6.5) and substituting the fitness values and the mean fitness, giving

$$\Delta q = \frac{pq[p(w_{Aa} - w_{AA}) + q(w_{aa} - w_{Aa})]}{\overline{w}}$$

$$= \frac{pq[p(1 - 1) + q(1 - s - 1)]}{1 - sq^2}$$

This gives

$$\Delta q = \frac{pq[-sq]}{1 - sq^2} = \frac{-spq^2}{1 - sq^2}$$

which is equation (6.8) in the text.

APPENDIX 6.3 DERIVATION OF FORMULAS FOR SELECTION AGAINST DOMINANT ALLELES

This section provides the derivation of equations (6.9)–(6.11).

In this model, the A allele is dominant and the a allele is recessive. Under selection against a dominant allele, the fitness values for the three genotypes are $w_{AA} = 1 - s, w_{Aa} = 1 - s$, and $w_{aa} = 1$. The mean fitness for a general model of natural selection is, from equation (6.1)

$$\overline{w} = p^2 w_{AA} + 2pq w_{Aa} + q^2 w_{aa}$$

Substituting the fitness values for selection against a dominant allele gives

$$\overline{w} = p^2(1 - s) + 2pq(1 - s) + q^2(1)$$

When multiplied out, this gives

$$\overline{w} = p^2 + 2pq + q^2 - sp^2 - 2spq$$

Because the first three terms add up to $1(p^2 + 2pq + q^2 = 1)$, the mean fitness is

$$\overline{w} = 1 - sp^2 - 2spq$$

Making use of the fact that $(p = 1 - q)$, this can be rewritten as

$$\overline{w} = 1 - s(1 - q)^2 - 2sq(1 - q)$$

which, when multiplied out, gives

$$\overline{w} = 1 - s(1 - 2q + q^2) - 2sq + 2sq^2$$
$$= 1 - s + 2sq - sq^2 - 2sq + 2sq^2$$
$$= 1 - s + sq^2$$

which is equation (6.9) in the text.

We now derive the formula for the allele frequency in the next generation (q') by substituting the fitness values into equation (6.2), which gives

$$p' = \frac{p(pw_{AA} + qw_{Aa})}{\overline{w}}$$
$$= \frac{p[p(1-s) + q(1-s)]}{\overline{w}}$$

Multiplying this out, and substituting the value for mean fitness from above, gives

$$p' = \frac{p[(p+q)(1-s)]}{1 - s + sq^2}$$

Because by definition $(p + q) = 1$, this equation reduces to

$$p' = \frac{p(1-s)}{1 - s + sq^2}$$

which is equation (6.10) in the text.

The amount of allele frequency change per generation ($\Delta p = p' - p$) is now derived using equation (6.4) and substituting the fitness values and the mean fitness, giving

$$\Delta p = \frac{pq[p(w_{AA} - w_{Aa}) + q(w_{Aa} - w_{aa}]}{\overline{w}}$$
$$= \frac{pq[p[(1-s) - (1-s)] + q(1 - s - 1)]}{1 - s + sq^2}$$

This gives

$$\Delta p = \frac{pq[p(0) + q(-s)]}{1 - s + sq^2} = \frac{-spq^2}{1 - s + sq^2}$$

which is equation (6.11) in the text.

APPENDIX 6.4 DERIVATION OF FORMULAS FOR SELECTION WITH CODOMINANT ALLELES

This section provides the derivation of equations (6.12)–(6.14).

In this model, the A and a alleles are codominant. Under selection against the codominant allele a, the fitness of the homozygote AA is highest, the fitness of the *aa* homozygote is lowest, and the fitness of the heterozygote is intermediate: $w_{AA} = 1, w_{Aa} = 1 - s/2$, and $w_{aa} = 1 - s$. The mean fitness for a general model of natural selection is, from equation (6.1), is

$$\overline{w} = p^2 w_{AA} + 2pq w_{Aa} + q^2 w_{aa}$$

Substituting the fitness values for selection against a dominant allele gives

$$\overline{w} = p^2(1) + 2pq \left(1 - \frac{s}{2}\right) + q^2(1 - s)$$

When multiplied out, this gives

$$\overline{w} = p^2 + 2pq + q^2 - \frac{2spq}{2} - sq^2$$
$$= p^2 + 2pq + q^2 - spq - sq^2$$

Because the first three terms add up to 1 ($p^2 + 2pq + q^2 = 1$), the mean fitness is

$$\overline{w} = 1 - spq - sq^2$$
$$= 1 - sq(p + q)$$

Making use of the fact that ($p + q = 1$), we can rewrite this as $\overline{w} = 1 - sq$, which is equation (6.12) in the text.

We now derive the formula for the allele frequency in the next generation (p') by substituting the fitness values into equation (6.3), which gives

$$q' = \frac{q[pw_{Aa} + qw_{aa}]}{\overline{w}}$$
$$= \frac{q\left[p\left(1 - \frac{s}{2}\right) + q(1 - s)\right]}{\overline{w}}$$

Multiplying this out, and substituting the value for mean fitness from above, gives

$$q' = \frac{q\left[p - \frac{sp}{2} + q - sq\right]}{1 - sq}$$
$$= \frac{q\left[p + q - s\left(\frac{p}{2} + q\right)\right]}{1 - sq}$$
$$= \frac{q\left[p + q - s\left(\frac{p + 2q}{2}\right)\right]}{1 - sq}$$

Because by definition ($p + q$) = 1, this equation reduces to

$$q' = \frac{q\left[1 - s\left(\frac{p + 2q}{2}\right)\right]}{1 - sq}$$

Now, since the term $(p + 2q)$ can be rewritten as $(p + q + q)$, which is equal to $(1 + q)$, the equation can be rewritten as

$$q' = \frac{q\left[1 - s\left(\frac{1+q}{2}\right)\right]}{1 - sq}$$
$$= \frac{q - sq(1+q)/2}{1 - sq}$$

which is equation (6.13) in the text.

The amount of allele frequency change per generation $(\Delta q = q' - q)$ is now derived using equation (6.5) and substituting the fitness values and the mean fitness, giving

$$\Delta q = \frac{pq[p(w_{Aa} - w_{AA}) + q(w_{aa} - w_{Aa})]}{\overline{w}}$$
$$= \frac{pq\left[p\left(1 - \frac{s}{2} - 1\right) + q\left(1 - s - \left(1 - \frac{s}{2}\right)\right)\right]}{1 - sq}$$

This gives

$$\Delta q = \frac{pq\left[p\left(-\frac{s}{2}\right) + q\left(-s + \frac{s}{2}\right)\right]}{1 - sq}$$
$$= \frac{pq\left[p\left(-\frac{s}{2}\right) + q\left(\frac{-2s + s}{2}\right)\right]}{1 - sq}$$
$$= \frac{pq\left[p\left(-\frac{s}{2}\right) + q\left(\frac{-s}{2}\right)\right]}{1 - sq}$$
$$= \frac{pq\left[-\frac{s}{2}(p + q)\right]}{1 - sq}$$

Because by definition $(p + q) = 1$, this equation reduces to

$$\Delta q = \frac{-spq/2}{1 - sq}$$

which is equation (6.14) in the text.

APPENDIX 6.5 DERIVATION OF FORMULAS FOR SELECTION AGAINST THE HETEROZYGOTE

This section provides the derivation of equations (6.15)–(6.17).

Under selection against the heterozygote, the fitness values of the three genotypes are $w_{AA} = 1, w_{Aa} = 1 - s$, and $w_{aa} = 1$. The mean fitness for a general model of natural selection is, from equation (6.1)

$$\overline{w} = p^2 w_{AA} + 2pq w_{Aa} + q^2 w_{aa}$$

Substituting the fitness values for selection against the heterozygote gives

$$\overline{w} = p^2(1) + 2pq(1 - s) + q^2(1)$$

When multiplied out, this gives

$$\overline{w} = p^2 + 2pq + q^2 - 2spq$$

Because the first three terms add up to 1 ($p^2 + 2pq + q^2 = 1$), the mean fitness is

$$\overline{w} = 1 - 2spq$$

which is equation (6.15) in the text.

We now derive the formula for the allele frequency in the next generation (q') by substituting the fitness values into equation (6.3), which gives

$$q' = \frac{q[pw_{Aa} + qw_{aa}]}{\overline{w}}$$

$$= \frac{q[p(1 - s) + q(1)]}{\overline{w}}$$

Multiplying this out, and substituting the value for mean fitness from above, gives

$$q' = \frac{q[p + q - sp]}{1 - 2spq}$$

Because by definition $(p + q) = 1$, this equation reduces to

$$q' = \frac{q[1 - sp]}{1 - 2spq}$$

$$= \frac{q - spq}{1 - 2spq}$$

which is equation (6.16) in the text.

The amount of allele frequency change per generation ($\Delta q = q' - q$) is now derived using equation (6.5) and substituting the fitness values and the mean fitness, giving

$$\Delta q = \frac{pq[p(w_{Aa} - w_{AA}) + q(w_{aa} - w_{Aa})]}{\overline{w}}$$

$$= \frac{pq[p(1 - s - 1) + q(1 - (1 - s))]}{1 - 2spq}$$

This gives

$$\Delta q = \frac{pq[p(-s) + q(s)]}{1 - 2spq}$$

$$= \frac{pq[-sp + sq]}{1 - 2spq}$$

$$= \frac{spq[-p + q]}{1 - 2spq}$$

$$= \frac{spq[q - p]}{1 - 2spq}$$

which is equation (6.17) in the text.

APPENDIX 6.6 DERIVATION OF FORMULAS FOR SELECTION FOR THE HETEROZYGOTE

This section provides the derivation of equations (6.18)–(6.20).

Under selection for the heterozygote, the fitness values of the three genotypes are $w_{AA} = 1 - s$, $w_{Aa} = 1$, and $w_{aa} = 1 - t$. The mean fitness for a general model of natural selection is, from equation (6.1)

$$\overline{w} = p^2 w_{AA} + 2pq w_{Aa} + q^2 w_{aa}$$

Substituting the fitness values for selection against the heterozygote gives

$$\overline{w} = p^2(1 - s) + 2pq(1) + q^2(1 - t)$$

When multiplied out, this gives

$$\overline{w} = p^2 + 2pq + q^2 - sp^2 - tq^2$$

Because the first three terms add up to 1 ($p^2 + 2pq + q^2 = 1$), the mean fitness is

$$\overline{w} = 1 - sp^2 - tq^2$$

which is equation (6.18) in the text.

We now derive the formula for the allele frequency in the next generation (q') by substituting the fitness values into equation (6.3), which gives

$$q' = \frac{q[pw_{Aa} + qw_{aa}]}{\overline{w}}$$

$$= \frac{q[p(1) + q(1 - t)]}{\overline{w}}$$

Multiplying this out, and substituting the value for mean fitness from above, gives

$$q' = \frac{q[p + q - tq]}{1 - sp^2 - tq^2}$$

Because by definition $(p + q) = 1$, this equation reduces to

$$q' = \frac{q[1 - tq]}{1 - sp^2 - tq^2}$$

$$= \frac{q - tq^2}{1 - sp^2 - tq^2}$$

which is equation (6.19) in the text.

The amount of allele frequency change per generation $(\Delta q = q' - q)$ is now derived using equation (6.5) and substituting the fitness values and the mean fitness, giving

$$\Delta q = \frac{pq[p(w_{Aa} - w_{AA}) + q(w_{aa} - w_{Aa})]}{\overline{w}}$$

$$= \frac{pq[p(1 - (1 - s)) + q(1 - t - 1)]}{1 - sp^2 - tq^2}$$

This gives

$$\Delta p = \frac{pq[p(s) + q(-t)]}{1 - sp^2 - tq^2}$$

$$= \frac{pq[sp - tq]}{1 - sp^2 - tq^2}$$

which is equation (6.20) in the text.

APPENDIX 6.7   CALCULUS-BASED DERIVATION OF THE EQUILIBRIUM ALLELE FREQUENCY UNDER SELECTION FOR THE HETEROZYGOTE

As shown throughout this chapter, we can think of natural selection in mathematical terms as a process that maximizes the mean fitness $(\overline{w})$ of the population. If you know differential calculus, it may have occurred to you that this focus on finding a maximum value of mean fitness lends itself easily to a calculus-based approach. If you do not know calculus, do not read any further!

Under selection for the heterozygote, the fitness values of the three genotypes are $w_{AA} = 1 - s, w_{Aa} = 1$, and $w_{aa} = 1 - t$. As shown in equation (6.18), the mean fitness under selection for the heterozygote is a function of the allele frequencies p and q, and the selection coefficients s and t:

$$\overline{w} = 1 - sp^2 - tq^2$$

By substituting $(1 - q)$ for p, we obtain

$$\bar{w} = 1 - s(1 - q)^2 - tq^2$$
$$= 1 - s(1 - 2q + q^2) - tq^2$$
$$= 1 - s + 2sq - sq^2 - tq^2$$
$$= 1 - s + 2sq - q^2(s + t)$$

We now have mean fitness written as a function of q. We then take the derivative of $\bar{w}$ with respect to q, giving

$$\frac{d\bar{w}}{dq} = 2s - 2q(s + t)$$

We can solve for the value of q associated with a maximum or minimum by setting this derivative equal to 0, which gives

$$q = \frac{s}{s + t}$$

Plotting $\bar{w}$ as a function of q shows that this value is a maximum.

APPENDIX 6.8 MUTATION–SELECTION EQUILIBRIUM UNDER SELECTION AGAINST A RECESSIVE ALLELE

Here, we consider a mutation from allele A to a recessive allele a at a mutation rate of μ. Under an equilibrium between mutation and selection, the change in allele frequency due to mutation (denoted Δq_M) and the change in allele frequency due to selection (denoted Δq_s) will cancel each other such that there is no *net* change, which means that

$$\Delta q_M + \Delta q_s = 0 \tag{A6.1}$$

The amount of change expected due to mutation is determined using the model of irreversible mutation in Chapter 4 [equation (4.1)]:

$$\Delta q_M = q' - q = (1 - p') - q$$
$$= 1 - p(1 - \mu) - q$$
$$= 1 - p + \mu p - q$$

Because $1 - p = q$, this means that

$$\Delta q_M = q + \mu p - q$$
$$= \mu p$$

The amount of change per generation due to selection against the recessive homozygote (Δq_s) has already been given in equation (6.8) in the chapter text. Substituting these values into equation (A6.1) gives

$$\Delta q_M + \Delta q_s = 0$$

$$\mu p - \frac{spq^2}{1 - sq^2} = 0$$

which means that

$$\mu p = \frac{spq^2}{1 - sq^2}$$

Factoring out p from both sides gives

$$\mu = \frac{sq^2}{1 - sq^2}$$

Now we can solve for q, which will give us the allele frequency of the mutant allele under a balance between irreversible mutation and selection against the recessive homozygote. Solving this equation gives

$$q^2 = \frac{\mu}{s(1 + \mu)}$$

and, after taking the square root of both sides, gives

$$q = \sqrt{\frac{\mu}{s(1 + \mu)}}$$

Because the mutation rate μ is a very low value, this means that $(1 + \mu)$ is approximately equal to 1, which, in turn, means that we can use the approximation

$$q \approx \sqrt{\frac{\mu}{s}}$$

APPENDIX 6.9 MUTATION–SELECTION EQUILIBRIUM UNDER SELECTION AGAINST A DOMINANT ALLELE

Here, we consider a mutation from allele a to a dominant allele A at mutation rate μ. We focus here on the frequency of the dominant allele, p. Under an equilibrium between mutation and selection, the change in allele frequency due to mutation (denoted Δp_M) and the change in allele frequency due to selection (denoted Δp_s) will cancel each other such that there is no *net* change, which means that

$$\Delta p_M + \Delta p_s = 0 \tag{A6.2}$$

Since we are looking at mutation of the a allele into the A allele, we modify equation (4.1) from Chapter 4 to express change in the a allele as

$$q' = q(1 - \mu)$$

The amount of change expected due to mutation is then determined using the model of irreversible mutation in Chapter 4 [equation (4.1)]:

$$\Delta_{p_M} = p' - p = (1 - q') - p$$
$$= 1 - q(1 - \mu) - p$$
$$= 1 - q + \mu q - p$$

Because $1 - q = p$, this means that

$$\Delta_{p_M} = p + \mu q - p$$
$$= \mu q$$

The amount of change per generation due to selection against the recessive homozygote (Δp_s) has already been given in equation (6.11) in the chapter text. Substituting these values into equation (A6.2) gives

$$\Delta p_M + \Delta p_s = 0$$
$$\mu q - \frac{spq^2}{1 - s + sq^2} = 0$$

which means that

$$\mu q = \frac{spq^2}{1 - s + sq^2}$$

Solving for q is difficult here, but an approximation is easy to do after noting that under this mutation model, the allele frequency q is close to 1. Using the approximation $q \approx 1$ reduces the equation to

$$\mu = sp$$

or

$$p = \frac{\mu}{s}$$

CHAPTER 7

NATURAL SELECTION IN HUMAN POPULATIONS

"Are we still evolving?" This is a question I hear quite often when telling people about my teaching and research interests in human evolution (not quite as often as the ever-popular "Where do you dig?," but still often enough). It is a good question, and one that continues to be debated among scientists today (Balter 2005). Some argue that because humans are biological organisms that are part of the natural world, we continue to be affected by the course of biological evolution. Others note that human *cultural* evolution occurs so much more rapidly than biological evolution that the latter has little effect on our future. For example, medical advances have reduced or eliminated many infectious diseases. New methods of agriculture have allowed more people to obtain more nutritious food. Is it not true that these, and other, cultural changes allowed more people to survive, thus countering or at least buffering natural selection to some extent? After all, life expectancy of a newborn child in the United States increased from 47 in 1900 to 77 in the year 2000 (Arias 2007). Children in developed nations are taller and mature earlier today than they did in the nineteenth century (Tanner 1990).

These rapid changes are clearly not the result of natural selection, but are instead changes in environmental conditions (primarily shifts in infectious disease and nutrition) that have allowed people to live longer and to reach their genetic potential for growth. Changes due to cultural adaptation appear to be more dramatic and rapid than through genetic adaptation. To some, such examples are evidence that the pace of our cultural evolution has essentially meant that we no longer evolve genetically in any significant way. On the other hand, it has also been argued that this view is a bit too simplistic. For one thing, the types of cultural changes leading to improvements in medicine and diet have not affected all human populations equally. The unfortunate truth is that many humans continue to live in impoverished environments without adequate healthcare or diet, and do not share the cultural buffers enjoyed by those in other parts of society or the world.

This is a fascinating debate, because if natural selection continues to shape human evolution, we will then want to know exactly how this occurs, and whether any predictions can be made as to our future. The primary purpose of this chapter is to present several case studies of natural selection, particularly those that show evidence of rapid evolutionary change in recent human evolution (i.e., the last 12,000 years since the initial beginnings of agriculture). These studies will provide perspective before returning to our opening question regarding current human evolution.

I. CASE STUDIES OF NATURAL SELECTION IN HUMAN POPULATIONS

The following case studies are not meant to represent everything that we know about natural selection in human populations, but instead are chosen to provide illustrative examples of different types of selection, different methods of analysis, and different lessons regarding natural selection.

A. Hemoglobin S and Malaria

The story of natural selection and the hemoglobin molecule is a classic in anthropology, providing an excellent example of balancing selection, rapid genetic change, and the effect of cultural and ecological influences on selection. Hemoglobin is a protein of the blood that transports oxygen to tissues throughout the body. The hemoglobin molecule of adults is made up of four protein chains. Two of these are identical and are called the *alpha chains*. The other two protein chains are the *beta chains*, and are identical to each other as well. The alpha chains are 141 amino acids in length, and the gene is on chromosome 16. The beta chains are 146 amino acids in length, and the gene is on chromosome 11 (Mielke et al. 2011). Our story here deals with the beta chain.

    The normal form of the beta hemoglobin gene is known as the A allele, and people with the AA genotype have hemoglobin that functions normally for transporting oxygen. A number of mutant alleles have been discovered, including the S, C, and E alleles, among other rarer forms (Livingstone 1967). Our focus here is on the S allele, also known as the *sickle cell allele*. This allele is due to a single mutation that replaces the sixth amino acid of the beta chain, glutamic acid with the amino acid valine. This small genetic chain has noticeable effect in individuals who carry two copies of the mutant gene; that is, they have the genotype SS. Those with the SS genotype have the genetic disease sickle cell anemia. In this case, low levels of oxygen can cause the red blood cells to become distorted, and change from their normal donut shape to the shape of a sickle (hence the name *sickle cell*). The deformed blood cells do not carry oxygen effectively, causing serious problems throughout the body's tissues and organs, and typically leading to death before adulthood without substantial medical intervention. Those who have the heterozygous genotype AS typically do not show this effect, and are known as *carriers*.

    In terms of models of natural selection, we can assign those with the AA genotype the highest relative fitness ($w = 1$). Heterozygotes have slightly lower fitness, and those with sickle cell anemia (SS) have the lowest fitness. Given this information, the models in Chapter 6 can be used to make some general predictions about the evolution and variation of the hemoglobin alleles (here, we focus on the A

and S alleles and ignore other alleles to make the points more clearly). Because the S allele is harmful in the homozygous case, we would expect selection against this allele. As S is a harmful mutant, we would expect its frequency to be very low, as expected under mutation–selection balance. Data from many human populations certainly fit this prediction. For example, the allele frequencies in a number of populations throughout the world are $A = 1$ and $S = 0$. In some other cases, S is not zero, but is very low, such as in Portugal ($S = 0.0005$), Libya ($S = 0.002$), and the Bantu of South Africa ($S = 0.0006$) (Roychoudhury and Nei 1988).

Such frequencies are consistent with a model of selection against an allele, and would simply be a good example of such selection if not for the fact that a number of human populations do *not* fit this general model. Some populations in the world have higher frequencies of the S allele that range from 0.01 to over 0.20 (Roychoudhury and Nei 1988). These frequencies are much higher than expected under a model of selection against the S allele. How could the frequency of a harmful allele be higher than expected? There are two possible answers. One is genetic drift. By chance, the frequencies of harmful alleles can drift upward given certain levels of population size and fitness values. The other explanation is balancing selection. When there is selection for the heterozygote, we have a situation where having only one copy of an allele actually confers higher fitness than someone with two copies, or someone with only one copy.

Balancing Selection and Hemoglobin S

It turns out that the higher frequencies of the S allele in a number of human populations can be explained by balancing selection. The clue to this effect is the geographic distribution of populations with higher (>0.01) frequencies of the S allele. Moderate to high frequencies of S are typically found in human populations in parts of west Africa and South Africa, as well as parts of the Middle East and India, among other regions. Further inspection shows that it is not a simple matter of geographic location—some African populations have high values, for example, while others have low values—but instead correlates with the distribution of epidemic malaria. Populations that have a history of malaria epidemics tend to have higher frequencies of the S allele.

Malaria is an infectious disease caused by a parasite, and is one of the most harmful diseases recorded in human history. Current estimates suggest that between 300 and 500 million cases of malaria occur each year, with between 1 million and 3 million deaths each year (Sachs and Malaney 2002). There are four different forms of malaria, each caused by a different parasite. The form of malaria relevant here is known as *falciparum malaria*, which is caused by the parasite *Plasmodium falciparum* and is the most fatal of the malarias. You cannot get malaria from someone else; the disease is transmitted by mosquitoes, a point that will be important shortly.

Your genotype for beta hemoglobin affects your susceptibility to falciparum malaria. Having an S allele renders red blood cells inhospitable to the malaria parasite, thus protecting the individual from its effects. If you have the normal hemoglobin genotype AA, you are more susceptible to malaria than someone who has an S allele. However, if you have two S alleles (genotype SS), you have sickle cell anemia, which is frequently fatal. On the other hand, if you have the heterozygous genotype AS, you do not have sickle cell anemia, and you are less likely to contract malaria. Thus, in a malarial environment, the heterozygote AS has the highest fitness followed by the AA genotype, and the SS genotype still has the lowest fitness.

We can see these fitness differences with an example provided by Bodmer and Cavalli-Sforza (1976 : 319) based on data from the Yoruba of Nigeria, a population in a malarial environment. The observed genotype numbers in adults were

$$AA = 9,365$$

$$AS = 2,993$$

$$SS = 29$$

(for a total of 12,387 adults). We use the allele counting method from Chapter 2 to see that there are 21,723 A alleles and 3,051 S alleles, for a total of 24,774 alleles. The allele frequencies are $A = \frac{21,723}{24,774} = 0.877$ and $S = \frac{3,051}{24,774} = 0.123$. Note that if this population were at Hardy–Weinberg equilibrium (and hence no selection), the expected genotype numbers would be computed by multiplying the expected Hardy–Weinberg proportions by the total number of adults (12,387), which would give

$$AA = p^2(12,387) = (0.877)^2(12,387) = 9527.2$$

$$AS = 2pq(12,387) = 2(0.877)(0.123)(12,387) = 2672.4$$

$$SS = q^2(12,387) = (0.123)^2(12,387) = 187.4$$

Comparing the expected numbers to the observed numbers, we see that there are fewer adults with genotype AA than expected, more with genotype AS than expected, and far fewer SS than expected. This population is clearly not at Hardy–Weinberg equilibrium. Because we know the biochemical relationship of hemoglobin alleles with malaria, and because there is a surplus of heterozygotes than expected, the evidence fits balancing selection.

Bodmer and Cavalli-Sforza (1976) derived the absolute fitness values for each genotype by taking the ratio of observed to expected numbers. This gives absolute fitness values of

$$AA = \frac{9,365}{9,527.2} = 0.983$$

$$AS = \frac{2,993}{2,672.4} = 1.120$$

$$SS = \frac{29}{187.4} = 0.155$$

We now transform the absolute fitness values into relative fitness values, which can be used in our model of balancing selection from Chapter 6, by dividing each fitness value by the highest fitness value. For balancing selection, this is the absolute fitness for the heterozygote, so we divide each absolute fitness value by 1.120, giving

$$AA = \frac{0.983}{1.120} = 0.878$$

$$AS = \frac{1.120}{1.120} = 1.000$$

$$SS = \frac{0.155}{1.120} = 0.138$$

These values mean that for every 100 people with genotype AS who survive to adulthood, roughly 88 with genotype AA survive and 14 people with genotype SS survive.

The selection coefficients for the homozygotes are obtained by subtracting the relative fitness values of each from 1, giving

$$s = 1 - w_{AA} = 1 - 0.878 = 0.122$$

$$t = 1 - w_{SS} = 1 - 0.138 = 0.862$$

We can now derive the expected equilibrium frequencies using equations (6.21) and (6.22) from Chapter 6 as

$$p = \frac{t}{s+t} = \frac{0.862}{0.122 + 0.862} = 0.876$$

$$q = \frac{s}{s+t} = \frac{0.122}{0.122 + 0.862} = 0.124$$

These values are almost identical with the observed allele frequencies, suggesting that this population has reached an equilibrium under balancing selection.

Culture Change and the Evolution of Hemoglobin S

The case of hemoglobin S provides a classic example of balancing selection (perhaps *the* classic example). It also provides a classic example of how culture affects genetic evolution in human populations. In this case, a scenario can be created depicting the frequency of S increasing in parts of Africa because of human populations practicing agriculture (Livingstone 1958; Bodmer and Cavalli-Sforza 1976). Before the introduction of agriculture, human populations living in heavily forested areas of western Africa would have experienced little problem with malaria because the mosquito species that spreads malaria do not thrive in such dense forest. Because there would be no selective advantage for heterozygotes, any S mutants would be selected against, and the frequency of S would be very low, as it is today in populations where malaria is not a problem.

The situation changed when horticulture (simple agriculture using hand tools) spread into the area. Forests were cleared for planting using slash-and-burn horticulture. The clearing of trees and subsequent changes to soil chemistry that reduced water absorption led to an increased in sunlit areas and pools of stagnant water, both of which are ideal conditions for spreading of mosquito infestations. As the mosquito population grew, they fed on human blood, and repeated bites allowed transmission of the malaria parasite. As malaria increased in humans, the balance between mutation and selection changed. Now those with the heterozygote had the highest fitness because they had resistance to malaria but did not suffer from sickle cell anemia. As the heterozygotes were selected for, the frequency of S would have increased. However, as the frequency of $S(=q)$ increases, so does the frequency of those with the SS homozygote ($=q^2$), who are then selected against, causing the increase in S to slow down and then stop to reach an equilibrium.

The scenario described above was modeled as shown in Figure 7.1. A single mutation (S) is introduced into a population of 1000 people, giving an initial allele frequency for S of 0.0005 ($=1$ in 2000 alleles, because each person has two alleles). The fitness values are those given above for the example from Bodmer and Cavalli-Sforza: $w_{AA} = 0.878, w_{AS} = 1.000$, and $w_{SS} = 0.138$. As shown in Figure 7.1, there is little initial change in S, as there are few heterozygotes. As the frequency of S continues to rise, a point is reached after about 20 generations where there are sufficient numbers of heterozygotes and the frequency of S increases rapidly. By around 60 generations, the higher frequency of S means that there are an increasing number of deaths due to sickle cell anemia, and the curve levels off to reach an equilibrium quickly. There is very little change after 60–70 generations. In terms of an average of 20–25 years per generation, 60–70 generations translates to between 1200 and 1750 years, which is a very high rate of evolutionary change. Although this model is somewhat simplistic, the results would be broadly consistent even taking factors such as repeated mutation and fluctuations due to genetic drift into account.

More recent research on the molecular genetics of the sickle cell mutation suggest that it arose and spread through selection and gene flow more than once, perhaps 5 times in different parts of Africa and Asia where malaria was common. We also need to look at the age of different mutations; the coalescent models described in Chapter 5 can be modified to estimate the time of an origin of a new

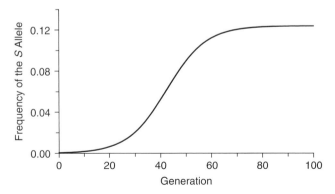

FIGURE 7.1 Simulation of the evolution of hemoglobin S in a malarial environment. The initial allele frequency of S was set at 0.0005 using fitness values of $w_{AA} = 0.878, w_{AS} = 1.000$, and $w_{SS} = 0.138$ (see text).

mutation. The origin of the sickle cell mutations dates to roughly 2000 to 3000 years ago (Mielke et al. 2011). Such research combined with what we know about fitness differences in hemoglobin genotypes shows that natural selection has taken place very quickly and in very recent (evolutionary) history. The idea that we stopped evolving when we reached the status of anatomically modern humans 200,000 years ago clearly does not apply to hemoglobin variants.

Another important lesson here is how cultural behavior (in this case, agriculture) changed the fitness values. What was most fit in one environment is not necessarily the most fit in another environment, something to remember given the rapidity with which we humans change our relationship to the environment. In the case of hemoglobin S, cultural change did not eliminate genetic change, but it did change its nature and trajectory, from selection against a mutation to a case of balancing selection. The situation could continue to change if efforts to eradicate malaria in tropical environments ultimately prove successful. In that case, the selective advantage for the heterozygote would disappear and selection against the S allele would *ultimately* reduce the allele frequency close to zero. However, as shown in Chapter 6, this could take a long time. For example, if we started with the equilibrium frequency of 0.124 for the S allele under balancing selection and changed the fitness values to $w_{AA} = 1.000, w_{AS} = 1.000$, and $w_{SS} = 0.138$, the frequency of S would still be 0.02 after 50 generations and 0.01 after 100 generations. Previous genetic variation does not disappear immediately even when culture changes the selective balance.

B. The Duffy Blood Group and Malaria

Given the intense effect of malaria on human mortality, it is not surprising that a number of genetic traits show evidence of natural selection because of malaria, including other hemoglobin alleles, genetic diseases known as *thalassemias*, and the enzyme glucose-6-phosphate dehydrogenase (G6PD). One such trait is the Duffy blood group, defined by the presence of different antigens on the surface of red blood cells. The gene for the Duffy blood group is found on chromosome 1 and consists of three codominant alleles: Fy^a, which codes for the a antigen; Fy^b, which codes for the b antigen; and Fy^0, which codes for the absence of any Duffy antigen. The Fy^0 allele, also known as the Duffy negative allele, is of particular interest here. The Duffy blood group is a receptor site for entry of certain types of malaria parasites into the body. Individuals who are homozygous for the Duffy negative allele cannot be infected by the parasite *Plasmodium vivax*, which causes vivax malaria (Chaudhuri et al. 1993). In populations that have experienced vivax malaria, we expect that there would be selection for the Duffy negative allele, ultimately leading to fixation of this allele.

The Duffy negative allele is found at highest frequencies in central and southern Africa, reaching a value of 1.0 (100%) in a number of populations. It is also found at moderate to high frequencies in northern Africa and the Middle East, at low frequencies in parts of India, and at or close to zero in native populations throughout the rest of the world (Roychoudhury and Nei 1988). This is the most extreme variation possible in a species, where an allele ranges from 0 to 1 in different populations, and such a range in values typically indicates selection across different environments. Because those homozygous for the Duffy negative allele cannot be infected by vivax malaria, we might expect a close relationship between the prevalence of vivax malaria and the frequency of the Duffy negative allele; that is, the longer vivax malaria has been around, the higher the frequency of the advantageous Duffy negative allele.

The actual geographic correspondence is almost exactly the opposite that we initially expect from a model of selection for an advantageous allele. Populations that have very high (or even fixed) frequencies of the Duffy negative allele tend to have little, if any, vivax malaria (Livingstone 1984). This might seem paradoxical at first, but keep in mind that when we survey genetic variation in the present day, we are seeing the net effect of a variety of factors in the past, which means that we need to consider time depth. For example, one possibility is that all of the selection for the Duffy negative allele took place in the past. Under this model, in populations experiencing vivax malaria, selection would favor those homozygous for the Duffy negative allele. Over time, the frequency of Duffy negative would increase, ultimately reaching a level at or near 100%. As this happened, however, more and more people would be immune to vivax malaria, and at some point, the disease would not be able to survive because of lack of susceptible individuals in the population to perpetuate the spread of the disease. At the end of this selection process, the human populations would have very high (or fixed) frequencies of the Duffy negative allele and there would be no more vivax malaria in the region, which is the situation we see today. An obvious lesson here is that we can see different relationships with selection depending on when we are observing the "present." Are we seeing the start, middle, or the end?

Further support for natural selection for the Duffy negative allele has come from molecular studies that have examined DNA sequences at and near the location of the Duffy blood group gene. Under natural selection, we expect that DNA variation should be reduced for the gene under selection (because other alleles are eliminated). Because of genetic linkage, where neighboring sections of DNA tend to be inherited together, we should also see reduced variation in DNA sequences near the gene, even if these neighboring sequences represent neutral genetic variation. Under natural selection, these neutral sequences are swept along in a process known as a **selective sweep**. Our ability to sequence sections of DNA allows us to compare observed levels of variation with those expected under a selective sweep and with those expected under neutral evolution (no selection). For the Duffy blood group and neighboring DNA sequences, the evidence shows reduced variation for those with the Duffy negative allele, confirming that selection has taken place (Hamblin and Di Rienzo 2000; Hamblin et al. 2002).

However, does this mean that the selection initially resulted from adaptation to vivax malaria? Not necessarily. Livingstone (1984) presents another hypothesis that can also explain the negative correlation between Duffy negative allele frequency and prevalence of vivax malaria. He suggests that there was already a high frequency of the Duffy negative allele in some populations, perhaps reflecting selection to some other disease. Consequently, these populations were already resistant to vivax malaria when it spread into Africa, but because almost everyone was immune, the disease never took hold. This alternative continues to be a possibility, although estimates of the date of the initial Duffy negative mutation from coalescent analysis suggest that it arose fairly recently, at about the same time as the origin of agriculture and the spread of malaria (Seixas et al. 2002). If so, then vivax malaria, and not some other disease, was responsible for fixation of the Duffy negative allele in Africa.

C.   The *CCR5-Δ32* Allele and Disease Resistance

The case study of the Duffy blood group raised the interesting possibility that current adaptive value may not reflect past selection. In the case of the Duffy blood group, the current adaptive value of the Duffy negative allele is that it protects against infection by the vivax malaria parasite. However, as pointed out by Livingstone (1984), that does not mean that this allele was due to adaptation to vivax malaria in the past. As noted above, the current evidence does suggest that vivax malaria *was* initially responsible for selection for the Duffy negative allele, but this correspondence between *initial* and *current* selection might not apply to other genes or traits. Evolutionary biologists have long argued this potential lack of correspondence when trying to explain the origin of an evolutionary trait. For example, the fact that birds use their wings to fly does not necessarily mean that wings *first* evolved for flight. Instead, they may have evolved for some other reason, such as thermoregulation or to reduce running speed, and only later were coopted for another purpose. As another example, consider upright walking in human evolution. Our ability to walk on two legs allows us to carry tools and weapons, but this does not mean that upright walking was *first* selected for because of this ability. Some other reasons may have been responsible at first, and only later was the ability to carry tools a significant factor.

On a microevolutionary level, we can see the same thing when looking at variation of the *CCR5-Δ32* mutation. The *CCR5* gene (short for C–C chemokine receptor type 5) is located on chromosome 3. This gene is responsible for the CCR5 protein, which functions in resistance to certain infectious diseases. A mutation of the *CCR5* gene results in the deletion of a 32-bp section of the DNA sequence of *CCR5*, and is known as the *CCR5-Δ32* (delta 32) mutation. The Δ32 mutation has a frequency of 0–14% in European populations, but is absent in the rest of the world (Stephens et al. 1998). Statistical analysis of DNA sequences near this locus provides strong support that the distribution of this allele has been shaped by natural selection (Bamshad et al. 2002).

As noted in Chapter 4, what makes the *CCR5-Δ32* allele interesting is the link to susceptibility to HIV, the virus that causes AIDS. Heterozygotes with one *CCR5Δ32* allele show partial resistance to HIV, and homozygotes show almost complete resistance to AIDS (Galvani and Slatkin 2003). However, AIDS has been around for only a short time in human history and therefore could not have been responsible for the initial elevation of this mutant allele in some European populations. Instead, higher frequencies of *CCR5-Δ32* most likely arose because of selection related to some *other* disease. What disease could have been responsible?

Answering this question is in part dependent on *when* the mutation first appeared in Europe. One coalescent analysis suggested a date of roughly 700 years ago (Stephens et al. 1998), which places the origin around the year 1300. One possibility that has been suggested is the pandemic of bubonic plague known as the *black death* that ravaged Europe from 1346 to 1352. However, Galvani and Slatkin (2003) modeled natural selection due to black death and found that even in a case where the mutant allele was dominant, an epidemic of bubonic plague would not have been strong enough to generate current levels of *CCR5-Δ32* in Europe. The period of potential selection due to bubonic plague was too short to result in frequencies much above 0.01. On the other hand, they found that models based on continuing epidemics of smallpox, another major historical disease, could account for current levels of *CCR5-Δ32*. Under a model of a dominant resistance allele, selection due to smallpox could have elevated the frequency of *CCR5-Δ32* to about 0.10 in 680 years. Assuming incomplete dominance extends the amount of time slightly, but still fits a relatively recent origin of the mutant allele.

Further insight into the evolution of the *CCR5-Δ32* allele comes from analysis of ancient DNA from human skeletal remains. Hummel et al. (2005) extracted DNA sequences from 14 skeletons from a mass grave of black death victims in Germany in 1350 and found that the allele frequency of *CCR5-Δ32*(= 0.142) was not significantly different from a control group of famine victims from Germany in 1316 that did not have the plague (allele frequency = 0.125). If individuals with *CCR5-Δ32* were less likely to be infected with bubonic plague, then the allele frequency of *CCR5-Δ32* in the black death mass grave should be significantly lower than in the control group, which is not the case.

Hummel et al. (2005) also found evidence for the *CCR5-Δ32* allele in 4 of 17 Bronze Age skeletons from Germany dating back 2900 years. These results show that the mutant allele was common in Europe over 2000 years before the black death. They conclude that bubonic plague was unlikely to have been a major factor in the evolution of *CCR5-Δ32*, and that smallpox is a more likely causal factor. It is possible, of course, that other infectious diseases might have also contributed to changes in *CCR5-Δ32* over time. In any event, this example shows again how adaptive relationships that we see today (in this case, AIDS resistance) cannot be used to explain the origin and evolution of a mutant allele.

D.   Lactase Persistence and the Evolution of Human Diet

The case studies presented thus far have focused on disease. Adaptation to disease through natural selection makes sense as disease directly affects one's probability of survival. As a change of pace, however, let us consider a different sort of selection: in this case, evidence of adaptation to changing diet. As with all mammalian species, human infants are nourished through breastfeeding. (The fact that in recent historical times some humans have used infant formula as an artificial substitute for mother's milk does not negate the fact that we have evolved, as have all mammalian species, to breastfeed.) Infant mammals produce the enzyme lactase, which allows lactose (milk sugar) to be broken down and digested. The typical pattern in mammals is for lactase production to shut down after the infant is weaned. After this, a mammal can no longer easily digest lactose.

Many humans today are lactose-intolerant, which means that they cannot produce the lactase enzyme after about 5 years of age. The physical effect of lactose intolerance can vary, but can include flatulence, cramps, bloating, and diarrhea. The interesting fact of human variation is that although many people are lactose intolerant, others have no trouble digesting lactose as they continue to produce the lactase enzyme throughout their lifetimes. The gene that controls lactase activity is on chromosome 2. Apart from some minor mutations that are found in low frequencies, there are two lactase activity alleles: *LCT*R*, which is the lactase restriction allele that codes for a shutoff of lactase production after weaning; and *LCT*P*, which is the lactase persistence allele that codes for continued lactase production. The lactase persistence allele is dominant, so individuals with one persistence allele (genotype *LCT*P/LCT*R*) or two persistence alleles (genotype *LCT*P/LCT*P*) can more easily digest lactose. The recessive homozygotes (*LCT*R/LCT*R*) are lactose-intolerant (Mielke et al. 2011).

Lactase persistence has an interesting geographic distribution. It is highest in northern Europe, and moderate in southern Europe and the Middle East. On average, lactase persistence is very low in African and Asian populations, although some African populations, such as the Fulani and the Tutsi, have moderate to high frequencies (Leonard 2000; Tishkoff et al. 2007). Variation in Africa is particularly revealing because some populations are found with high frequencies of lactase persistence, and some are found with very low levels (and thus have a high prevalence of people who are lactose-intolerant).

The critical factor that explains global variation in lactase persistence is diet. Populations that have a history of dairy farming tend to have higher frequencies of the lactase persistence allele. Natural selection has been proposed as an explanation, where the lactase persistence allele was selected for in dairying populations because of the nutritional advantage among those that were able to digest milk (higher levels of fats, carbohydrates, and proteins, in addition to extra water from the milk). Because the cattle are domesticated, there was likely also selection for cattle that produced more milk and better nutritional content, a good example of the process of coevolution of two species. A study of European cattle revealed that the geographic distribution of six milk protein genes in cattle was correlated with levels of lactase persistence in humans and the locations of prehistoric sites associated with the early adoption of cattle farming (Beja-Pereira et al. 2003). As humans adapted to a diet that included milk, they also selected cattle that provided better quantity and quality milk.

The high frequency of the lactase persistence allele in Europe and in some African populations poses an interesting problem in understanding the origin and spread of the allele. Did it arise in Europe and spread to Africa, or arise in Africa and then spread to Europe? On the other hand, did it evolve independently in both Europe and in Africa? Molecular genetic analysis of the lactase persistence allele shows that the third hypothesis is correct. The specific mutation responsible for lactase persistence in Europeans is a change from the base C to the base T at base pair position 13910 in the lactase persistence gene. However, this variant is rare or absent in African populations, which instead undergo a change from G to C at base pair position 14010. Thus, different mutations appeared in different places that produced lactase persistence, and each mutation was selected for independently when dairy farming arose and milk became an important part of the human diet (Tishkoff et al. 2007).

The estimated age of lactase persistence mutations fits the archaeological evidence for the origin of domestication of cattle. Cattle farming began in northern

Africa and the Middle East between 7500 and 9,000 years ago, which fits with the estimated age of 8000–9000 years for the T-13910 mutation in Europeans. A somewhat earlier age estimate of 2700–6800 years for the C-14010 mutation in Africa fits the younger age of 3300–4500 years ago for cattle domestication in sub-Saharan Africa (Tishkoff et al. 2007).

It is clear from both the European and African evidence that the increase in lactase persistence occurred in a very short time in an evolutionary sense. All of these changes, genetic and cultural, took place only within the last 10,000 years, which is a high degree of evolutionary change. Tishkoff et al. (2007) estimate a selection coefficient associated with lactase restriction of between $s = 0.035$ and $s = 0.097$. We know from Chapter 6 that natural selection can be rapid for a dominant allele. For example, selection for a dominant allele starting at a frequency of 0.001 using a selection coefficient of $s = 0.05$ can result in an allele frequency of 0.13 after 100 generations, 0.71 after 200 generations, and 0.87 after 300 generations. In terms of human lifetimes, 300 generations is about 7500 years using a length of 25 years for each generation. We can play with the numbers, but it is clear that a lot of evolution is possible in 10,000 years!

E. Genetic Adaptation to High-Altitude Populations

The examples presented thus far have focused on adaption through natural selection to disease and diet. There are also examples of how humans have adapted to different physical environments as our ancestors expanded out of Africa and spread across the world. One particular challenge has been dealing with the physiologic stress of living in a high-altitude environment. Some humans have adapted to living permanently at heights in excess of 2500 ms (~8200 ft, or 1.6 mi) above sea level (Beall and Steegmann 2000). Some high-altitude populations live as high as 5400 m above sea level (Beall 2007). The main stress of high-altitude environments is hypoxia, which is a shortage of oxygen. As altitude increases, barometric pressure decreases. Consequently, there is less oxygen available in the blood, and arterial oxygen saturation drops, causing severe physiologic stress (Frisancho 1993).

When a low-altitude native enters a high-altitude environment, the hypoxic stress can be countered to some extent by a variety of physiologic mechanisms, such as increased red blood cell production and increased respiration. Over time, these stresses can often be quite serious. Some adaptation occurs in infants and children who move to high altitude through a process known as *developmental acclimatization*, where there are changes in the body during the growth process when adapting to an environmental stress. For example, children born at low altitudes who have moved to high altitude show an increase in aerobic capacity compared with those that stay at low altitude. Further, the younger the child who moves to high altitude, the greater the change (Frisancho 1993).

A number of studies have shown that children in high-altitude populations around the world tend to have increased chest dimensions, indicating greater lung volume relative to body size (Frisancho and Baker 1970; Frisancho 1993). Much of this growth appears to be the result of developmental acclimatization, as the growth response to high altitude is related to the age at which a child moves there; the earlier the age, the greater the response. However, there are exceptions to this general trend, which suggest that the growth response is not always solely developmental acclimatization, and there may be genetic differences as well (Greksa 1996; Beall and Steegmann 2000).

If there are at least some genetic influences on high-altitude environment, then some physiologic and biochemical traits may have been shaped by natural selection. More recent genetic studies have provided evidence of natural selection to high altitude, along with an interesting lesson about the nature of natural selection itself, which is that different populations have adapted in different ways to the hypoxic stress of high altitude (Beall 2007; Storz 2010). Comparisons of two widely studied high-altitude populations have provided some interesting contrasts. One group is the native inhabitants of the Andean Plateau in South America, descended from human populations moving into the region about 11,000 years ago. The other group is the native inhabitants of the Tibetan Plateau, who colonized the area about 25,000 years ago. Unlike low-altitude natives that visit high altitudes and suffer from hypoxic stress, members of these high-altitude populations have normal levels of oxygen consumption, although the physiologic pathways are different among Tibetans and Andeans. For example, an elevation in ventilation is a typical response of a low-altitude native to the stress of high altitudes. This response is also seen in Tibetans, but not in Andeans, who have lower resting ventilation levels. Other differences between Tibetans and Andeans include the level of oxygen in the blood, the level of oxygen saturation in the blood, and levels of hemoglobin concentration (Beall 2007). Work has begun to identify genes that are responsible for high-altitude adaptations and how they have been selected for in Tibetan populations (e.g., Beall et al. 2010; Simonson et al. 2010; Yi et al. 2010).

Current (as of 2011) research suggests that both Andean and Tibetan populations have evolved different genetic adaptations to high-altitude stress. This suggests that the starting point for both populations, in terms of initial genetic variation, was different, and that natural selection provided alternative means of adapting. The important lesson here is that there may be different paths of selection that operate differently depending on the types of genetic variation available, which, in turn, is influenced by variation in mutation and genetic drift.

Apart from the evidence for multiple adaptive solutions, studies of the genetics of high-altitude adaptation provide yet another example of the rapid pace of recent human evolution, because the high-altitude populations have been in those environments only within recent evolutionary history (i.e., within the past 11,000–25,000 years, which again is a short time ago in evolutionary history).

F. The Evolution of Human Skin Color

The final example of natural selection in this chapter deals with another example of human genetic adaptation to different environmental conditions. Human skin color (pigmentation) is a quantitative trait that shows an immense amount of variation between human groups around the world, ranging from some whose average level of pigmentation is very dark to those who are extremely light in color. Further, there are no apparent breaks in this distribution. Humans do not come in a set number of discrete shades; that is, humans are not made up of "black," "brown," and "white" people, but instead come in every shade along a continuum from very dark to very light (Relethford 2009).

The wide range of variation in human skin color is one clue that it has been affected in the past by natural selection, and the specific geographic distribution of skin color is another. For native human populations (those who have not recently migrated), skin color tends to be darkest at or near the equator, and decreases with increasing distance from the equator, both north and south. This relationship is shown clearly in Figure 7.2, which plots the relationship between skin color (as typically measured as the percentage of light reflected off the skin at a wavelength of 685 nm) and distance to the equator for 107 human populations in the Old World (Africa, Europe, Asia, Australasia). Although there is some scatter, in general indigenous human populations show a strong correlation of skin color and distance from the equator. Such striking correspondence with the physical environment is a clue for natural selection, an inference strengthened by the fact that the amount of ultraviolet (UV) radiation received also varies by distance from the equator. UV radiation is strongest at the equator and diminishes with increasing distance away from the equator. Skin color, levels of UV radiation, and distance from the equator are all highly correlated (Jablonski and Chaplin 2000). The obvious conclusion is that the evolution of human skin color has been linked to varying degrees of UV radiation. Because pigmented skin acts as a protective barrier against UV radiation, dark skin color has evolved as a protection against excessive UV radiation in areas at or near the equator. Farther away from the

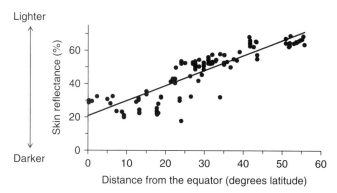

FIGURE 7.2 Geographic distribution of human skin color in the Old World. The dots represent samples of 107 human populations in the Old World that were measured using an E.E.L. reflectometer that measures the percentage of light at 685 nm reflected off of human skin on the upper inner arm, which is relatively unexposed in many human societies. Because skin color varies between the sexes, only male samples are included here as there are more studies of male skin color than female skin color). Distance from the equator is taken as the absolute value of the latitude for each population. Data sources for 102 samples are listed in Relethford (1997) and supplemented with data on five Australian native samples listed in Relethford (2000). The solid line is a linear regression showing the best fit of a straight line to the data points.

equator, where levels of UV radiation are lower, the problem is not excessive, but insufficient, UV radiation, and lighter skin color likely evolved to allow more UV radiation to affect human skin. Given these geographic relations, we can turn to examining some possible selective forces that are related to UV radiation.

The evolution of human skin color can be examined in more detail by studying the geographic distribution of skin color in the world today in conjunction with evidence from the fossil and archaeological records. Given the evidence that our ancestors evolved first in Africa and then spread elsewhere, we need to consider two separate, but related, questions: (1) why dark skin evolved in populations at or near the equator and (2) why did humans moving farther from the equator evolved lighter skin.

The Evolution of Dark Skin

Human ancestors first evolved in Africa, and it was only 2 million years ago that the species *Homo erectus* (a species ancestral to us) became the first hominin species to disperse outside of Africa. The oldest evidence of this species is from central eastern Africa, where it was adapted to living to the open grasslands of the savanna environment. Anthropologists suspect that loss of hair and the evolution of efficient sweat glands occurred in this environment, where exposure to solar heat would have been a major stress, selecting for more and improved sweat glands and the loss of hair, which increases the rate of heat convection (Bramble and Lieberman 2004). Comparative studies of African apes indicate that our early ancestors before this time probably had lightly pigmented skin (Jablonski and Chaplin 2000). As they lost their protective fur through adapting to the heat, those individuals with darker pigmentation would have been at a selective advantage because darker skin protects from the damage caused by UV radiation. What were the specific selective pressures? Some have suggested protection against skin cancer (Robins 1991), but it is not clear how much of a selective force this would have been given that most skin cancers affect older individuals who have already reproduced, and thus these cancers would have minimal impact on fitness (Jablonski and Chaplin 2000). Severe sunburn is a possible factor, as it could lead to skin infection and can damage sweat glands, which would be hazardous in a hot environment (Robins 1991). Jablonksi and Chaplin (2000) argue that one of the more likely negative consequences of UV radiation damage would be the photodestruction of folate, a needed nutrient. Low levels of folate affect the development of the embryonic neural tube and can reduce sperm production, both of which could affect fitness. Another hypothesis has been suggested by Elias et al. (2010), who propose that darker skin evolved to resist infections, in line with evidence of how skin functions as a barrier.

Of course, it is also possible that some combination (or all) of the above factors contributed to lower fitness for those with light skin and higher fitness for those with dark skin. The debate might be better expressed in terms of the relative contribution of differences in adaptive value to overall fitness. In terms of the models of selection for quantitative traits described in Chapter 6, the evolution of dark skin in our ancestors is an example of directional selection; with each generation, there would be a shift toward darker average skin color.

Molecular genetic analysis gives us a clue when the initial selection for dark skin began in our ancestors. Rogers et al. (2004) examined variation at the *MC1R* (melanocortin 1 receptor) locus, a gene that influenced human skin color. They examined patterns of diversity in this gene and compared it to levels of diversity expected after a rapid selective sweep for dark skin in Africa and subsequent accumulation of mutations after the sweep. The idea here is that when selection takes place, genetic diversity at the locus affected is reduced to zero. Afterward, mutations will accumulate, allowing us to estimate the time since selection (given a number of assumptions about mutation rate and population size). Rogers et al. conclude that the earliest possible date for selection at the *MC1R* gene took place 560,000 years ago, and the most reasonable estimate of population size suggests a date of 1.2 million years ago. Placing the rapid selection of dark skin in Africa at about a million years ago suggests that hair loss and subsequent selection for dark skin occurred in the ancestral species *Homo erectus*, a species that anatomical evidence suggests was the first to move about long distances on the African savanna. In addition, the suggested timing of the loss of body hair agrees with the anatomical evidence suggesting that characteristically human long-distance endurance running appeared about 2 million years ago (Bramble and Lieberman 2004). It seems likely that the evolution of improved sweating and loss of body hair, an adaptation to more frequent exposure to solar heat, was followed by a rapid evolution of dark skin to replace the protective function of hair.

The Evolution of Light Skin

The fossil and archaeological evidence shows that populations of *Homo erectus* left Africa and moved into Eurasia roughly 1.8–1.6 million years ago, reaching Southeast Asia and the easternmost edge of Europe, and later as far north as China. Later species, including *Homo heidelbergensis* and *Homo sapiens*, also occupied many northern parts of Eurasia, where present-day populations show light skin. The steady decline in pigmentation with latitude (Figure 7.2) is often considered evidence of past selection for lighter skin color and against darker skin color. Many suggestions for the specific nature of this selection focus on the fact that UV radiation also declines with distance from the equator. Thus, the dangers from excessive UV radiation exposure that selected for darker skin color in equatorial human populations diminished as humans moved away from the equator. Our ancestors no longer needed to have such dark skin.

However, there must be more to the explanation than a reduction in UV radiation, because although our ancestors did not *need* to have dark skin, there would be no reason why they could not have it. The distribution of human skin color shows, however, that evolutionary change *did* occur, and to understand why, we need to consider the possible evolutionary *disadvantage* of having dark skin and/or the evolutionary *advantage* of having light skin farther from the equator.

One favored explanation for light skin in regions of reduced UV radiation relates to vitamin D, an essential nutrient. Vitamin D deficiency can result in poor bone growth and other medical problems such as immune system abnormalities (Wagner et al. 2008). Although many of us receive vitamin D today from vitamin supplements or milk (to which vitamin D has been added during processing), both are relatively recent cultural developments. The primary source of vitamin

D for our ancestors would have been the sun, because UV radiation stimulates a biochemical synthesis of vitamin D where 7-dehydrocholesterol is converted into the substance previtamin D, which is then converted into vitamin D. Pigmented skin reduces the amount of vitamin D synthesis because it blocks UV radiation. As humans moved into northern climates, the amount of UV radiation declined, not only reducing the problems of UV damage but also increasing the problem of insufficient vitamin D synthesis. Natural selection would then favor those with lighter skin through directional selection. Once an optimal level of pigmentation was reached, it would then be maintained through stabilizing selection. The optimal level of pigmentation would depend on the relative influences of problems associated with excessive or inadequate UV radiation. As humans moved north, it would be necessary to compensate for reduced levels of UV radiation and vitamin D synthesis by being lighter on average.

Not everyone has accepted the vitamin D hypothesis for the origin of light skin. Robins (1991, 2009) argues that the bone disease rickets, one of the proposed dangers from insufficient vitamin D synthesis, is a recent development associated with pollution due to industrialization, and was not a risk factor in the more distant past. Robins also argues that even though vitamin D synthesis would be less in dark-skinned people in northern latitudes, it would still be sufficient for most metabolic needs. Chaplin and Jablonski (2009) disagree with Robins' interpretations of rickets and point out that there are other health problems associated with insufficient vitamin D levels.

Some have argued that the critical environmental change was not the level of vitamin D synthesis, but instead changes in temperature associated with changes in latitude. This "cold injury hypothesis" rests on suggested evidence that darker skin is more likely to suffer from frostbite, such that light skin would be favored in northern latitudes that are colder than at the equator (Post et al. 1975; Robins 1991). Beall and Steegmann (2000) argue that the evidence for increased cold injury in those with dark skin is spurious, and they favor the vitamin D hypothesis.

Some have suggested an entirely different type of selection to explain the evolution of light skin: **sexual selection**. This idea, first proposed by Darwin, is an explanation for the evolution of traits through competition among members of the same sex for mates or preferences for mating with those that have certain characteristics, such as bright feathers in some species of birds. Either way, the result would be selection for certain traits in the reproductive process. Aoki (2002) has suggested that the evolution of light skin in humans is in part due to past sexual selection, based on the assumption that most human males prefer to mate with females with lighter skin color. If true, then the geographic distribution of human skin color reflects the balance between selection for dark skin in populations at or near the equator and the (presumed) universal preference for choosing light-skinned mates. As humans moved further from the equator, the strength of selection for dark skin diminished and the strength of selection for lighter skin increased, leading to the lightest skin in northern latitudes. A major problem with the sexual selection hypothesis is that the assumption of mating preference for light skin existed in the past and is not a product of recent cultural change (Mielke et al. 2011). Another problem with the model is that, if it is correct, we would expect the differences between male and female skin color to increase in populations farther from the equator as selection for darker skin diminishes. However, Madrigal and Kelly (2007) tested this hypothesis using global data on skin color and found that this was not the case.

The vitamin D hypothesis, the cold injury hypothesis, and the sexual selection hypothesis all have one thing in common—they all agree that the evolution of light skin was due to some type of natural selection! Not everyone agrees. Some have suggested that the evolution of light skin might reflect the relaxation of selection for dark skin and neutral evolution (mutation and genetic drift). Although some interpretations of the molecular evidence are consistent with this idea (e.g., Harding et al. 2000), others differ and provide evidence of past selection (e.g., Rogers et al. 2004; Norton et al. 2007). In fact, a study of patterns of variation in six different pigmentation genes shows that different alleles have been selected for in Europeans and East Asians (Norton et al. 2007). As with lactase persistence, there might have been multiple paths of selection for light skin in human populations.

Advances in the genetic technology for reconstruction of ancient DNA sequences have also revealed some interesting findings regarding the Neandertals of Europe, a group of early humans who apparently interbred with modern humans (see Chapter 9). Lalueza-Fox et al. (2007) sequenced fragments of one of the human pigmentation genes from the fossil remains of two European Neandertals, and found that their alleles indicated that they likely had pale skin and red hair. Given the geographic location of the Neandertals, light skin is not unexpected. However, these sequences revealed a different mutation in the skin color gene than is found in living Europeans, suggesting the independent evolution of light skin in Neandertals. This study provides further support for multiple pathways for natural selection.

II.   ARE HUMANS STILL EVOLVING?

Given the case studies of natural selection in human populations presented above, we can now return to the questions about the recent, current, and future evolution of humans. Many of the examples presented in this chapter have shown how quickly natural selection can operate, and show that the variation we see in the world today resulted from recent natural selection. Much of the variation we see in genetic adaptations to malaria, such as the spread of hemoglobin *S*, have arisen since the origin and spread of agriculture, which is a recent event in human prehistory, beginning about 12,000 years ago and more recently in different parts of the world. The spread of lactase persistence, linked with the spread of dairy farming, is another example of an evolutionarily recent event that took place roughly in the past 4000–9000 years. Genetic adaptations such as the *CCR5-Δ32* allele are also likely to date to the time since the origin of agriculture, as epidemic infectious diseases were more of a problem in the larger, sedentary agricultural populations than in the smaller populations that were characteristic of our hunting–gathering past prior to the advent of agriculture. Indeed, it is likely that the changing demographic, ecological, and dietary changes that accompanied the origin and spread of agriculture increased the opportunity for human genetic evolution. The idea that our species made the transition from hunting and gathering to agriculture with no genetic impact does not hold up to the available evidence. There have certainly been major genetic changes at a number of loci that date to selection in our recent past.

Although we have strong evidence for selection in the recent past, there remains the question of how typical selection has been. As noted in Chapter 5, many argue that a considerable part of our species' genetic diversity reflects a neutral process reflecting an equilibrium with genetic drift. For many years now, there has been considerable debate over whether human genetic diversity can best be explained by a "neutralist" perspective, a "selectionist" perspective, or some combination of these. In the context of the case studies presented here, we have to ask whether such examples of recent selection represent the exception or the rule. What evidence exists for additional selection that either has occurred in the recent past or continues today?

A. How Do We Detect Recent Selection?

It is, in fact, difficult to study natural selection in many populations, including humans. Our best examples, such as hemoglobin S or skin color, involve major differences in fitness of different genotypes and phenotypes. Smaller differences in fitness, although evolutionarily relevant, would be more difficult to detect. In addition, small differences in fitness might be swamped by other factors, such as random genetic drift.

Because of technological advances, we are now collecting large amounts of data from the entire human genome, which have allowed some new ways of exploring the question of recent natural selection. A new field, sometimes referred to as **population genomics**, has developed to formulate evolutionary tests and inferences based on the entire human genome. Such studies consist of examining statistical measures of population-genetic diversity across the entire genome and comparing the distribution of these statistics to a theoretical distribution derived under the assumption of no selection (a neutral model). Differences of the observed distribution from the expected distribution can provide clues that natural selection has occurred, although not necessarily any detail about the specific action of selection.

Akey (2010) identifies three different types of studies that have relied on genomewide data to look for evidence of recent natural selection. One method looks at the **site frequency spectrum**, which is a plot of genetic variation (such as allele frequencies) that shows how common different values are across the genome, and compares these distributions with those expected under a neutral model. When natural selection favors a particular allele, it becomes more common in the population over time (as shown in Chapter 6). As noted earlier, neutral sections of DNA that are close by on the same chromosome and are linked to the allele that are selected for are caught up in a selective sweep and also become more common (a process also known as "hitchhiking"). Selection for a particular allele leads to reduced diversity for that locus, because one allele is being favored over others. Any neutral loci that are linked to the locus of selection will also show this effect, and will show reduced variation relative to a neutral expectation. Thus, when we compare an observed site frequency spectrum with the theoretical expectation under neutrality and see more low-frequency alleles than expected, we have evidence for selection. Of course, this is not always easy, because similar patterns can be mimicked by demographic changes, such as recent population growth (something quite common in recent human evolution).

Another clue to possible selection uses measures that reflect the amount of allele frequency differences between populations in different environments. Large differences are typical of past natural selection as populations adapt to different environments, such as the examples we have seen when comparing malarial and nonmalarial environments (hemoglobin *S*) and northern and equatorial regions (skin color). Under neutrality, we expect to see a distribution where most loci show small differences and only a small number that have larger differences (by chance). If we compare the observed distribution with the theoretical distribution, and find more high-frequency differences than expected, then we have suggested evidence for natural selection.

Linkage disequilibrium (LD) is also used to search for natural selection. Recall from Chapter 2 that some haplotypes will occur more or less often than expected when they are in linkage disequilibrium. Rapid evolution during a selective sweep is one way that this can happen, when insufficient time has elapsed to allow recombination to shuffle the newly selected mutant across other haplotypes. Further, the amount of linkage disequilibrium under selection will depend on the distance (along the chromosome) from the site of selection. A pattern of decay in linkage disequilibrium with increasing genomic distance is a characteristic sign of a recent selective sweep (Hawks et al. 2007). Each of these methods has some problems (including the inability to distinguish selection from demographic changes such as population growth) (Akey 2010).

More recently, a number of studies have used variants of the above tests on large databases on human genetic diversity, mostly single-nucleotide polymorphisms (SNPs). Some examples include analyses of several million SNPs using linkage disequilibrium (e.g., Wang et al. 2006; Hawks et al. 2007) and geographic divergence measures (e.g., Coop et al. 2009). These studies suggest that natural selection has been very important in recent human evolution. In particular, it appears that the shift of humans from hunting and gathering to agriculture, which started about 12,000 years ago, has been accompanied by a great deal of natural selection. The demographic and biological consequences of the initial spread of agriculture, including increased population size, sedentary lifestyle, epidemics of infectious disease, and reduced dietary diversity, are well known from studies of the skeletal biology of early agriculturists (e.g., Larsen 2000). Genomic studies have shown that this rapid cultural change has also had an impact on our genetic evolution. Instead of cultural evolution negating genetic evolution, we are finding evidence of how cultural change has accelerated genetic evolution. It is noteworthy that the regions of DNA that show evidence of recent selection are related to resistance to infectious disease and metabolism (Hawks et al. 2007; Akey 2010), because disease and diet were two aspects of human biology that were most affected by the agricultural revolution.

Although such studies offer promise for finding examples of recent natural selection, more work needs to be done to pinpoint the specific mechanisms of selection for individual loci. Although a number of studies have found many regions of DNA that are statistically suggestive of recent selection, only a small number of actual genes have been linked to selection (Gibbons 2010). Another complication is the problem that demographic shifts, such as population growth, can mimic in part the genetic patterns expected under a selective sweep, which means that further work is needed to explore selection under a range of demographic models (Akey 2010). Further, some analyses show a complex pattern of selection interacting with genetic drift and gene flow. For example, Coop et al (2009) notes that there is little evidence of a selective sweep moving a mutant allele to fixation, and that more selection has consisted of "partial sweeps" characterized by varying selective pressures over time as well as fluctuations due to genetic drift. Work is also underway to examine how selection for some traits might involve more than one gene (Gibbons 2010). Further work is also needed to determine how prevalent these kinds of positive selection have been compared with the typical pattern of selection eliminating harmful mutations, or what Weiss and Buchanan (2009) call "failure of the frail."

B.   The Future?

Although the issue of how much selection has taken place in our recent evolutionary history, and the strength of such selection, remains debated, it is clear that there is evidence of natural selection in humans since the spread of agriculture. Will natural selection continue in our species, and if so, what direction might it take? There are some reasonable predictions regarding the first question, whereas the second is more an exercise in speculative science fiction than making direct inferences from population genetics theory. The last 12,000 years of human history have seen a shift from hunting and gathering to agriculture to civilization, followed by the industrial revolution and a postindustrial world with a growing global economy. Our population size has increased at a phenomenal rate. Prior to the development of agriculture, the maximum number of people who could survive by hunting and gathering is estimated to be about 6 million (Weiss 1984), whereas the world population had reached 1 billion by 1850 (Weeks 2005), and by 2010, roughly 7 billion (Population Reference Bureau 2010). All of these (and other) demographic and cultural changes could influence the future direction of natural selection.

The dramatic increase in size of the human species has been cited as a potential influence on future natural selection. For one thing, the larger the population, the greater the number of new mutations with each generation, as shown in Chapter 4. The fate of new mutants is also affected by population size. In small populations, many new mutations will be lost quickly because of genetic drift, as shown in Chapter 5. This will occur even if the new mutation has the potential to be adaptive, because in a small population the influence of drift generally has greater impact than the selective advantage. For large populations, drift has less impact, and a new mutant is less likely to become extinct quickly through drift, thus giving an advantageous mutant a chance to reach a higher frequency. In addition, the greater number of mutations in a larger population means that the same mutation might soon occur again, and could be reintroduced in the next generation even if it was initially lost through drift (Reed and Aquardo 2006).

As pointed out by Hawks et al. (2007), the rapid cultural evolution in our recent past, particularly the origin of agriculture, did not supplant genetic evolution but instead created many new opportunities for new genetic evolution. As our species continues to change culturally, we should expect this situation to continue. However, the speed at which cultural evolution occurs is still faster than that of genetic change; the minimum amount of time for genetic evolution is a generation, whereas cultural change can occur instantaneously (consider, for example, the rapid cultural, demographic, and economic shifts that you have seen within your lifetime). In terms of possible natural selection, this means that in some cases the type of assumptions we made in Chapter 6 about constancy of fitness values over dozens of generations would no longer apply. Consider, for example, a case of selection for an allele (A) where having two alleles is better than having one, and where having one allele is better than having none. If we assign rather dramatic differences in fitness to the genotypes ($w_{AA} = 1, w_{Aa} = 0.75, w_{aa} = 0.5$) and start with an initial allele frequency for A of 0.1, selection will cause the allele frequency to increase to 0.42 after five generations [this was computed using equation (6.13)]. This is indeed very rapid selection. If this locus affected genetic susceptibility to a disease, would this mean that we would necessarily see a rapid evolution of resistance to the disease? Perhaps, although five generations is about 100–125 years of time, which is a *very* long time for cultural evolution these days. Maybe a cure (or vaccine) for the disease would be found during this time.

Although it seems clear that the potential for new genetic variations has never been greater for the human species, it is also clear that the cultural and environmental landscape is also changing very rapidly. An adaptive allele in the present might soon become neutral, or the reverse, depending on the direction of change within which human evolution takes place. Although we can discuss some generalities, *specific* predictions of future human genetic evolution may not be possible because of the rapid pace, and uncertain direction, of future cultural change. For example, who can predict when a new cure will be discovered for a disease, or the evolution of a new disease, under changing environmental conditions? How can we predict new sources of food? To take the questions further, what lies in the future of human evolution in terms of genetic engineering, nanotechnology, space colonization, or any number of possible futures that are now part of science fiction? Such ideas may seem bizarre, but history has a way of showing yesterday's science fiction becoming today's fact. (For example, we used to joke in graduate school about solving questions about Neandertals by getting their DNA. What was science fiction then is established fact today.)

In sum, advances in molecular genetics suggest that the case studies of natural selection presented in this chapter are not exceptions, and there may be a great deal of "recent" human genetic evolution. In addition, we are likely to continue to have the potential for future genetic change, with more newly introduced mutations being introduced each generation than at any other time in history. However, while it is clear that cultural evolution has not halted genetic evolution, it is also clear that cultural evolution also continues, and at a greater potential rate than genetic evolution. Future human evolution will continue to be a complex interplay of genetic and cultural change.

III. SUMMARY

Studies of human genetic and morphological variation have allowed us to understand a number of significant examples of natural selection in "recent" human evolution, with many examples showing genetic adaptation within only the past 12,000 years or so. One of the classic examples of natural selection in human populations is the change in allele frequencies of hemoglobin *S*, which is selected for in malarial environments. Here, the higher fitness of the heterozygote (*AS*) relative to the *AA* genotype (susceptibility to malaria) and the *SS* genotype (presence of has sickle cell anemia), is also a classic example of balancing selection.

Other genetic studies also show the importance of infectious disease in recent natural selection. The fixation of the Duffy negative allele is a genetic adaptation related to the spread of vivax malaria. The association of the *Δ32* mutation of the *CCR5* gene with resistance to AIDS is interesting, but the evolutionary history of this allele suggests an initial adaptation to a different disease, perhaps smallpox.

A number of examples of natural selection are related to other aspects of human evolution, including changes in diet. The rapid increase in the lactase persistence allele, allowing those possessing this allele to digest milk sugar even as adults, occurred independently in different human populations as an adaptation for increased nutrition available from dairy farming, a recent cultural innovation. Other genetic changes reflect the propensity of our ancestors to adapt to different environments in all corners of the world. For example, humans that have moved to high-altitude environments have adapted genetically to the physiologic stress of lowered amounts of oxygen, although in different ways. A striking example of environmental adaptation is the wide distribution in human skin color, reflecting different stresses associated with different levels of ultraviolet radiation.

More recent studies of the human genome suggest that such examples are not rare, and that our DNA contains clues to a considerable amount of natural selection within the past 12,000 years. The rapid demographic and cultural changes of the human species also suggest different patterns of genetic diversity and potential selection from those of our recent ancestors. Human evolution has not ceased, but continues, both genetically and culturally, in a complex manner.

CHAPTER 8

GENE FLOW

In 1908, A Russian immigrant to the United States, Israel Zangwill, staged a play entitled *The Melting Pot* that likened the assimilation of immigrants in American culture to the melting of metal alloys in a large crucible (pot). Since then, the term has been used and debated in the context of discussions of culture contact and the extent to which cultures are assimilated. Contact between different human populations has always existed, and can range from local contact between neighboring groups to long-distance movements of groups. In the last 500 years or so, the pace of human contact has further increased. Such contact, both small and large in scale, and local and distant, has many social, political, economic, demographic, and ecological implications. Contact between human groups also has genetic implications, where the movement of genetic material from one population to another (gene flow) can cause changes in allele frequencies. On a broader level, gene flow can be seen as the evolutionary glue holding populations in a species together, and reduction or elimination of gene flow is necessary to initiate the process of speciation.

This chapter includes a brief review of the basic concept and simple models of gene flow, applicable to a wide range of organisms. In addition, certain models used to analyze human gene flow are given particular focus, because it is often easier to assess gene flow in humans than in other organisms. I often see squirrels outside from my office window, but it is not easy to determine what part of campus they might have come from. Some probably live in the trees right across the road, but for all I know some might be migrants from farther away on campus or from someplace in the local neighborhood. How could I tell? In some studies of animal migration, they capture animals, tag them, and then recapture them later. Migration is much easier to assess in humans. Whereas we need to track squirrels, we can actually ask humans where they and their parents originated.

Although we sometimes talk about *migration* and *gene flow* as synonymous, for many contexts we may need to draw a distinction. In humans, we typically treat migration as a change in residence as compared with a brief visit. For example, if I take a vacation trip to the Caribbean, we would not regard that as an actual migration, although if I moved to the Caribbean, that would constitute migration. The line between the two can sometimes be complicated; for example, a college

Human Population Genetics, First Edition. John H. Relethford.
© 2012 Wiley-Blackwell. Published 2012 by John Wiley & Sons, Inc.

student might live away from home most of the year, but this might not be considered a migration in many contexts. Gene flow, on the other hand, involves the actual movement of genes (and DNA sequences) as the result of physically moving, whether temporarily or permanently. In this context, we can consider gene flow as what happens when an organism moves and reproduces in a different location, regardless of whether it remains there. In some cases, we might treat migration and gene flow as the same, but in other cases, the two terms may have different meanings.

I.   THE EVOLUTIONARY IMPACT OF GENE FLOW

One assumption of Hardy–Weinberg equilibrium is that the population remains closed to other populations; that is, there is no gene flow. In fact, throughout our discussions of mutation, drift, and selection, we have always focused on a single population. In the real world, however, the existence of a single-population species is very rare. For example, humans live in thousands of populations around the world, which are all interconnected by varying levels of gene flow in the present and the past. What is the evolutionary impact of gene flow?

A.   Introducing New Alleles

Thus far, the only way that we have seen a new allele enter a population is through mutation; drift and selection can increase or decrease the frequency of a new mutation, but they cannot bring about a new allele. Although mutation is the ultimate source of all new alleles, a new mutant allele can be introduced into a different population through gene flow. Picture two populations, A and B, where everyone has two copies of a specific allele. Now, imagine a mutation occurring in population A. Although it is possible for the same mutation to occur in population B, it is highly unlikely to happen in the same generation. On the other hand, if someone from population A carries the mutant allele and moves to population B and reproduces, the mutant allele is now established in population B through gene flow. Gene flow allows the spread of new mutants throughout a species, subject in each population to the further effects of drift and selection (we can see how complex microevolution can get when considering all four evolutionary forces at the same time).

B.   Reducing Genetic Differences between Populations

The main impact of gene flow is to reduce genetic differences between populations. Here, we can picture gene flow as analogous to mixing paint. Imagine starting with two gallon cans of paint, one with red paint and one with white paint. Take a cup of paint from the red can and mix it into the white can at the same time as taking a cup of white paint and mixing it into the red can. After mixing, the can of red paint is slightly lighter and the can of white paint is slightly pinker. If you repeat the mixing, the color of the paint in the two cans will become increasingly similar. After enough cups of paint have been swapped and mixed, you will have two identical cans of pink paint.

We can picture gene flow in an analogous manner where the alleles of two or more populations are mixed together. Although alleles are not paint, and do not actually merge together, the point here is that the mixing of gene pools can alter the allele frequencies. As an example, imagine two populations (A and B), where the frequency of a given allele is $p = 1.0$ in population A and $p = 0.0$ in population B. Now, imagine that 5% of the individuals in population A move into population B and reproduce there, while 5% of the individuals in population B move into population A and reproduce there. In other words, the two populations exchange 5% of their genes. This means that 95% of the individuals in population A remain in population A. The allele frequency in population A a generation later is made up of the mixing of 95% from population A with an allele frequency of $p = 1.0$ and 5% from population B with an allele frequency of $p = 0.0$. This gives the new allele frequency in population A of

$$(0.95)(1.0) + (0.05)(0.0) = 0.95$$

At the same time, we can figure out the allele frequency in population B a generation later by noting that it will consist of 5% from population A with an allele frequency of $p = 1.0$ and 95% from population B with an allele frequency of $p = 0.0$, giving

$$(0.05)(1.0) + (0.95)(0.0) = 0.05$$

Because of gene flow, the allele frequencies in populations A and B have changed from 1.0 and 0.0 to 0.95 and 0.05. They are still quite different, but closer than they were initially. Over time, gene flow will make the two populations increasingly similar. To figure out the allele frequencies in the next generation, we use the same mixing proportions (95% and 5%) and use the new allele frequencies for populations A and B ($p = 0.95$ and $p = 0.05$). Thus, the allele frequency in population A after two generations of gene flow is

$$(0.95)(0.95) + (0.05)(0.05) = 0.905$$

and the allele frequency in population B is

$$(0.05)(0.95) + (0.95)(0.05) = 0.095$$

If we go to the third generation, we simply use these new allele frequencies with the same amount of mixing to get allele frequencies of

$$(0.95)(0.905) + (0.05)(0.095) = 0.8645$$

for population A, and

$$(0.05)(0.905) + (0.95)(0.095) = 0.1355$$

for population B. We can repeat the same process many times from one generation to the next to see the long-term effect of gene flow. Figure 8.1 shows the results

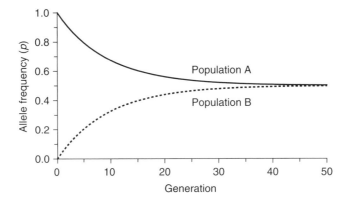

FIGURE 8.1 Simulation of gene flow between two populations. Populations A and B start with allele frequencies of $p = 1.0$ and $q = 0.0$, respectively. With each generation, 5% of each population is exchanged with the other, leading to a reduction of the allele frequency differences over time.

of gene flow for 50 generations. We see that gene flow quickly reduces the difference between the two populations such that they are almost identical after 40 generations.

II. MODELS OF GENE FLOW

Now that we have seen the basic effect of gene flow to reduce genetic differences between populations, we can examine several different models of gene flow in more specific detail. As with previous treatment of evolutionary forces, we will consider gene flow by itself to start with and then add complexity by examining the interaction with other evolutionary forces.

A. The Island Model

The simplest place to start with understanding how gene flow works is to examine the case of one-way migration as shown in the **island model** (or more specifically, the continent–island model, as there are other types of island models used in population genetics). Here, we imagine an island that receives a certain amount of migration from the mainland, but this migration is strictly in one direction—no migration occurs from the island to the mainland. In this way, we can see what effect gene flow from the mainland has on the allele frequencies of the island. We use the symbol m to illustrate the migration rate, which is the proportion of alleles in the island that come from the mainland each generation. Because m represents the genetic contribution from the mainland, this means that the proportion $(1 - m)$ of the island's alleles comes from the island (because the two proportions, one from the mainland and one from the island, have to add up to 1). The dynamics of this model are shown in Figure 8.2.

In order to simulate the one-way gene flow under the island model, we need to know the initial allele frequencies of the island and the mainland. Here, we

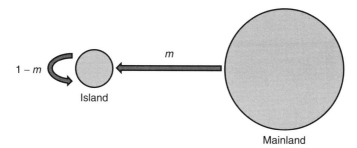

FIGURE 8.2   The island model. Gene flow is one-way from a large mainland population into a smaller island, such that the allele frequency in the mainland affects the allele frequency in the island, but not the other way around. m represents the migration rate each generation and is the proportion of alleles in the island's next generation that comes from the mainland. The quantity $(1 - m)$ represents the proportion of alleles in the island's next generation that comes from the island. For example, a value of $m = 0.1$ indicates that 10% of the alleles in the island come from the mainland, and $(1 - m) = 0.9$ indicates that 90% of the alleles on the island come from the island in the previous generation.

use p_0 to represent the initial allele frequency (in generation 0) of the island. We use the symbol P (uppercase p) to represent the allele frequency of the mainland. Because our model only involves gene flow, and since there is no gene flow from the island to the mainland, this means that the mainland allele frequency P stays the same from one generation to the next. In order to calculate the allele frequency on the island in the next generation, we multiply the proportions of alleles from the island and the mainland by their allele frequencies, respectively. For the first generation, this will be

$$p_1 = (1 - m)p_0 + mP$$

where m is the migration rate (i.e., the proportion of alleles from the mainland). Because P (and hence mP) does not change, we can also predict the allele frequency on the island after an additional generation of one-way gene flow by substituting p_1 for p_0 in the above equation, giving

$$p_2 = (1 - m)p_1 + mP$$

Because this is an iterative equation of the type encountered in Chapter 4, we can use the method in Appendix 4.1 to derive the allele frequency in any given generation, p_t as

$$p_t = (1 - m)^t(p_0 - P) + P \qquad (8.1)$$

(see Appendix 8.1 for the derivation).

The effect of one-way gene flow in the island model is shown in Figure 8.3 using an initial allele frequency for the island of $p_0 = 0.8$, a constant allele frequency on the mainland of $P = 0.2$, and a per generation migration rate of $m = 0.05$. Over time, the allele frequency on the island becomes increasingly similar to that on the

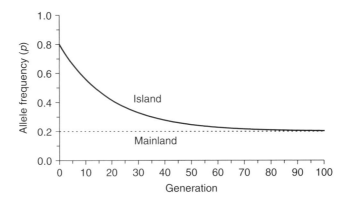

FIGURE 8.3   Change in allele frequencies over time in the island model. The allele frequencies in each generation are computed using equation (8.1). The initial allele frequency on the island is $p_0 = 0.8$. The allele frequency on the mainland, which is constant over time, is $P = 0.2$. The amount of one-way gene flow per generation (the migration rate) is $m = 0.05$. Over time, the island will become increasingly similar to the mainland.

mainland, showing that the genetic makeup of the island is eventually replaced by that of the mainland.

B. Two-Way Gene Flow

The island model focuses on gene flow in one direction, and allele frequency does not change in the source population (mainland). Although this model fits some cases, a model that allows gene flow in two directions is more applicable to many situations. Figure 8.4 shows a simple model of two-way gene flow where the rate of migration is the same in both directions. Here, two populations, A and B, exchange migrants and genes at a rate of m per generation. This number represents the proportion of alleles in one population from the other population. Each population receives a proportion $(1 - m)$ of its alleles from itself.

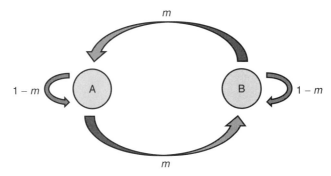

FIGURE 8.4   Two-way gene flow. Two populations, A and B, each have a proportion m of alleles from the other population and a proportion $(1 - m)$ of alleles from themselves.

We start with the rate of gene flow per generation m and allele frequencies for population A and B in generation t, which we label as p_{A_t} and p_{B_t} (the first subscript refers to the population and the second subscript, to the generation). We can estimate the allele frequency in population A in the next generation $(t + 1)$ by multiplying the allele frequencies of each population by the proportion of alleles provided to population A:

$$p_{A_{t+1}} = (1 - m)p_{A_t} + mp_{B_t} \qquad (8.2)$$

Likewise, the allele frequency in population B in the next generation is derived using the same logic, such that

$$p_{B_{t+1}} = mp_{A_t} + (1 - m)p_{B_t} \qquad (8.3)$$

As an example, consider initial allele frequencies of $p_{A_0} = 1.0$ and $p_{B_0} = 0.0$ and a gene flow rate of $m = 0.1$ per generation. After one generation of gene flow, the allele frequency is population A is computed using equation (8.2), giving

$$p_{A_1} = (1 - 0.1)(1.0) + (0.1)(0.0) = 0.9$$

and the allele frequency in population B is computed using equation (8.3), giving

$$p_{B_1} = (0.1)(1.0) + (1 - 0.1)(0.0) = 0.1$$

Note that the allele frequencies in the two populations are closer than before. We can then extend gene flow to subsequent generations by taking these new allele frequencies and substituting them back into equations (8.2) and (8.3), giving allele frequencies in the second generation of

$$p_{A_2} = (1 - 0.1)(0.9) + (0.1)(0.1) = 0.82$$

and

$$p_{B_2} = (0.1)(0.9) + (1 - 0.1)(0.1) = 0.18$$

This process can be extended to additional generations as shown in Figure 8.5, which shows how the allele frequencies become increasingly similar to each other. In fact, by about 20 generations, the allele frequencies are essentially the same. Note that the curves resemble those seen in Figure 8.1, which was also modeled using two-way gene flow. When you compare Figures 8.1 and 8.5, you will see that convergence of the allele frequencies is faster in Figure 8.5. This is because the rate of gene flow used to construct Figure 8.5 ($m = 0.1$) is greater than that used in Figure 8.1 ($m = 0.05$).

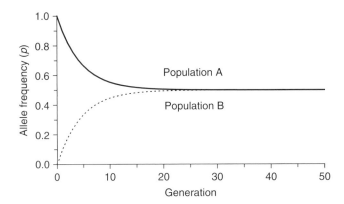

FIGURE 8.5   Two-way gene flow with initial allele frequencies of $p_A = 1.0$ and $p_B = 0.0$ and a rate of gene flow of $m = 0.1$. Compare the speed at which the allele frequencies converge with that of Figure 8.1, where a lower rate of gene flow ($m = 0.05$) was used.

C.   Kin-Structured Migration

The examples using two-way gene flow show clearly that gene flow acts to make populations more similar to each other over time, and that the higher the rate of gene flow, the more rapidly this convergence occurs. Gene flow is most often considered a homogenizing force that reduces the genetic difference between populations. However, this is technically correct only if we make the hitherto unstated assumption that the individuals that migrate and reproduce are a random sample of the source population. This may not always be the case. In some situations, the migrants may be related. One way that migrants can be related occurs in some small-scale human societies when part of a population splits off and then fuses with another population. There are other examples of entire families moving into a new population. When the migrants are related, we call this **kin-structured migration**.

Anthropologist Alan Fix has looked at the genetic effects of kin-structured migration on genetic differences between populations (Fix 1978, 1999). Gene flow usually acts to reduce differences between populations, and kin-structured migration can slow this process down, or even reverse it in some cases, leading to *increased* genetic differences between groups. Fix (1978) looked at kin-structured migration among the Semai Senoi, a swidden (slash-and-burn) agricultural group in the Malayan Peninsula in Malaysia. Over time, populations split and the resulting groups of migrants were (and are still) typically kin. Using simulations incorporating genetic drift and gene flow based on the demography of the population, Fix found that although kin-structured migration flow would still reduce genetic differences between Semai Senoi populations over time, the actual reduction would be less than if the migrants were simply a random sample. Although kin-structured migration is not universal, it has been found in other human populations, including the Yanomama Indians of South America as well as the founding populations of Tristan da Cunha and Plymouth in the United States (Fix 1999).

III. GENE FLOW AND GENETIC DRIFT

Of course, gene flow does not operate alone, and to get a better idea of the genetic effects of gene flow, we need to consider how it interacts with genetic drift. To do this, we need to consider in more detail the concept of **between-group variation** introduced in the first chapter. In previous chapters, we have considered the effect of microevolution on genetic variation (measured by heterozygosity or a similar statistic) *within* a single population. This **within-group variation** refers to genetic differences (variation) between individuals within a single population. Between-group variation (also referred to as *among-group variation*) looks at differences between the genetic compositions of two or more populations. For example, if two populations each have allele frequencies of $p = 0.6$ and $q = 0.4$, there is no genetic difference *between* these populations, although there is variation *within* both populations.

On average, genetic drift increases genetic differences between populations. The ultimate fate of genetic drift is the fixation or extinction of an allele. Because this is a random process, over time different alleles will become fixed in different populations, thus increasing the genetic variation between populations. On the other hand, gene flow counters drift and reduces genetic differences between populations. Clearly, the two evolutionary forces act in opposition to each other, and we need to see how and when they might reach an equilibrium between the amount of between-group variation added by genetic drift and the amount lost by gene flow.

A. Measuring Genetic Variation between Populations

Before looking at the equilibrium between genetic drift and gene flow, we need to consider briefly how we can measure between-group variation. In Chapter 5, when we considered the balance between mutation and genetic drift, we looked at heterozygosity (H) as a measure of the amount of genetic variation *within* a population. Here, we consider another measure (F_{ST}) as a measure of genetic variation *between* populations. Picture a region made up of a number of populations. Examples include a group of villages on a Pacific island, a group of rural towns in the English countryside, or a group of different ethnic neighborhoods in a city, among many other possible examples. Although we will consider real examples from human populations later in this chapter, for the moment we will approach this idea in a more abstract manner by considering a number of subpopulations that make up the total population. Figure 8.6 shows a graphic illustration where the total population (T) is made up of three subpopulations (S). This type of model refers to a **hierarchical population structure** where the subpopulations are nested inside the total population.

For our purpose here, we can look at genetic variation in three different ways:

1. The amount of genetic variation in the total population
2. The amount of genetic variation within the subpopulations
3. The amount of genetic variation between the subpopulations

Note that quantities 2 and 3 make up the total amount of variation:

Total variation = (variation within groups) + (variation between groups)

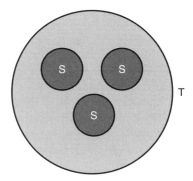

FIGURE 8.6 Hierarchical population structure. The total population (T) is made up of three subpopulations (S).

When we look at the effects of gene flow and drift (or other factors) on genetic differences between groups, we are interested in what proportion of the total variation is due to genetic variation. This proportion, denoted F_{ST}, is

$$F_{ST} = \frac{\text{variation between groups}}{\text{total variation}}$$

Because total variation is made up of the sum of within-group variation and between-group variation, we can express the above equation in terms of total and within-group variation as

$$F_{ST} = \frac{(\text{total variation}) - (\text{variation within groups})}{\text{total variation}} \tag{8.4}$$

This measure looks at variation in the subpopulations (S) relative to the total (T) population.

A simple example will help clarify these concepts. Imagine a single locus with two alleles, A and a, with the following allele frequencies in three subpopulations that have equal population sizes:

Subpopulation 1: $p = 0.6, q = 0.4$
Subpopulation 2: $p = 0.52, q = 0.48$
Subpopulation 3: $p = 0.5, q = 0.5$

We also need to know the allele frequencies in the *total* population, pooling all three subpopulations into a single group. Because the three populations all have equal population sizes, we can simply take the average allele frequencies, designated as $\bar{p}$ and $\bar{q}$, as

$$\bar{p} = \frac{0.6 + 0.52 + 0.5}{3} = 0.54$$

$$\bar{q} = \frac{0.4 + 0.48 + 0.5}{3} = 0.46$$

If the subpopulations had different population sizes, then we would have computed an average weighted by population size.

The first thing we are interested in looking at is the amount of genetic variation within the *total* population. This quantity is the heterozygosity based on the allele frequencies of the total population ($\bar{p}$ and $\bar{q}$), and is computed as shown in Chapter 5 for two alleles as

$$H_T = 2\bar{p}\,\bar{q} = 2(0.54)(0.46) = 0.4968$$

Note that we use the subscript 'T' to designate that this refers to the total population. The next quantity of interest is the amount of genetic variation *within* the subpopulations. Here, we simply take the heterozygosity for each subpopulation and then average them. Using the formula for heterozygosity for two alleles ($H = 2pq$), we obtain heterozygosity values of $H = 0.48$ for subpopulation 1, $H = 0.4992$ for subpopulation 2, and $H = 0.5$ for subpopulation 3. The average of these three values gives us an amount of genetic variation *within* groups of

$$H_W = \frac{0.48 + 0.4992 + 0.5}{3} = 0.4931$$

Here, we use the subscript W to designate variation *within* subpopulations.

We now return to equation (8.4) and see that we have measures for both total variation (H_T) and within-group variation (H_W), and can derive F_{ST} as

$$F_{ST} = \frac{\text{(total variation)} - \text{(variation within groups)}}{\text{total variation}}$$

$$= \frac{H_T - H_W}{H_T} \tag{8.5}$$

For our simple example, $H_T = 0.4968$ and $H_W = 0.4931$, giving an F_{ST} value of 0.0074. As this is a proportion, we can state that 0.74% of the total genetic variation is due to variation between groups, leaving 99.26% of the total genetic variation due to variation within groups.

The F_{ST} values in human populations frequently range from close to zero to higher values between 0.05 and 0.10 (within 5–10%) (Jorde 1980). As will be outlined in more detail below, genetic drift increases F_{ST} and gene flow decreases F_{ST}. A population with a moderate F_{ST} is often characterized by relatively small populations and/or low rates of gene flow. Large populations and/or high rates of gene flow tend to produce smaller F_{ST} values.

B. Equilibrium between Gene Flow and Genetic Drift

You might be wondering where the symbol F in F_{ST} comes from. We have seen other examples of F earlier in the book, including the concept of identity by descent under inbreeding (Chapter 3) and the probability of identity by descent due to genetic drift (Chapter 5). The use of the same letter is not accidental; all of these uses refer to various types of a **fixation index**, which is a measure of the proportional reduction in heterozygosity relative to Hardy–Weinberg conditions. Equation (8.5) shows the proportional reduction that occurs when a population is structured into subpopulations (S) rather than consisting of one large population (T) within which all individuals have an equal chance of mating. Whenever a population is subdivided in this way, there will be some reduction in heterozygosity, and F_{ST} provides an index of the degree of reduction.

Because we are dealing with the probability of identity by descent, we can extend models developed earlier to the case of an interaction of gene flow and genetic drift. Like mutation, gene flow acts to reduce the probability of identity by descent—an allele introduced from outside a population means that there will be no common ancestry within the population. When looking at gene flow, we define m as the rate of gene flow per generation, which is interpreted as the probability that an allele comes from outside the population. In practical terms, we typically estimate m in human populations as the proportion of individuals in a population that were born in another population. Identity by descent is possible only when both alleles have *not* come into the population due to gene flow. In Chapter 5, we dealt with a similar idea when considering the interaction between mutation and drift, which means that the logic and derivation of the probability of identity by descent under gene flow and drift can be derived by substituting the rate of gene flow (m) for the rate of mutation (μ) in Appendix 5.2. Thus, equation (A5.6) becomes

$$F_{ST_t} = (1 - m)^2 \left[\frac{1}{2N} + \left(1 - \frac{1}{2N}\right) F_{ST_{t-1}} \right] \tag{8.6}$$

Figure 8.7 shows values of F_{ST} over time for a population size of $N = 100$ and a migration rate of $m = 0.05$. F_{ST} increases rapidly at first because of the cumulative effects of genetic drift. Over time, gene flow moderates this increase, until an equilibrium of between-group variation is reached. As before, N as used here refers to the breeding population size in an idealized model, and in actual analysis we would want to consider the effective population size.

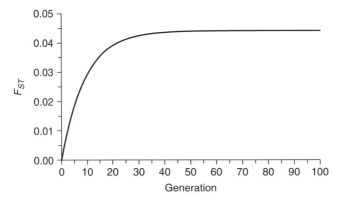

FIGURE 8.7   Change in F_{ST} over time for $N = 100$ and $m = 0.05$. The F_{ST} values were computed using equation (8.6) starting with an initial value of $F_{ST} = 0$. Initially F_{ST} increases rapidly as a result of genetic drift, but then levels off because of gene flow to approach an equilibrium of roughly $F_{ST} = 0.044$.

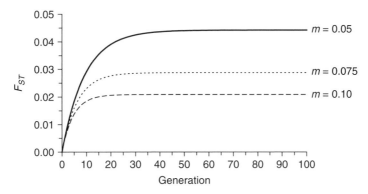

FIGURE 8.8   Change in F_{ST} over time for different values of migration rate. In each case, the population size was set to $N = 100$. The F_{ST} values were computed using equation (8.6) starting with an initial value of $F_{ST} = 0$. The larger the migration rate, the smaller the equilibrium value of F_{ST}.

The eventual equilibrium between genetic drift and gene flow depends on both population size and migration rate. Figure 8.8 shows three examples of changes in F_{ST} over time where the population size is held constant at $N = 100$ but the migration rate varies. The higher the migration rate, the lower the equilibrium value of F_{ST}. Figure 8.9 shows three examples of change in F_{ST} over time where the migration rate is held constant at $m = 0.05$ but the population size varies. The larger the population size, the smaller the equilibrium value of F_{ST}.

An approximate estimate of the equilibrium value of F_{ST} can be obtained using the same method as used for the equilibrium between mutation and genetic drift shown in Appendix 5.2, but substituting migration rate for mutation rate in equation (A5.8). Doing this gives an approximate equilibrium value for F_{ST} of

$$F_{ST} \approx \frac{1}{1 + 4Nm} \tag{8.7}$$

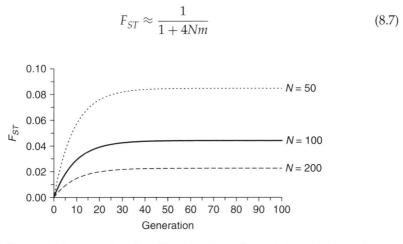

FIGURE 8.9   Change in F_{ST} over time for different values of population size. In each case, the migration rate was set to $m = 0.05$. The F_{ST} values were computed using equation (8.6) starting with an initial value of $F_{ST} = 0$. The larger the population size, the smaller the equilibrium value of F_{ST}.

Note what the term Nm in equation (8.7) actually means. It is the product of the population size and the migration rate, which means that Nm is the number of migrants each generation. Let $M = Nm$ be the number of migrants into any of the subpopulations for each generation, and equation (8.7) simplifies to

$$F_{ST} \approx \frac{1}{1 + 4M} \qquad (8.8)$$

This means that the equilibrium rate of F_{ST} depends only on the *number* of migrants For example, the case where $N = 100$ and $m = 0.03$ will have the same number of migrants ($M = 3$) as will the case where $N = 50$ and $m = 0.06$, or the case where $N = 1000$ and $m = 0.003$. In each case, the approximate F_{ST} at equilibrium is 0.0769. Another important point is that very few migrants are needed to reduce F_{ST} to the levels seen in most human populations. Figure 8.10 shows the relationship between the number of migrants and the approximate equilibrium value of F_{ST}. A change from only $M = 1$ to $M = 2$ changes the equilibrium value of F_{ST} from 0.2 to 0.11, and a further change to $M = 3$ changes the equilibrium value of F_{ST} to 0.08. Many studies of human populations show F_{ST} values between 0.01 and 0.05, which translates to values of M ranging from about 25 to 5, assuming that equilibrium has been reached. Note that these figures are approximations based on an assumption of an infinite (or at least a very large) number of subpopulations; more complex formulas are used to deal with a small number of populations (Rogers and Harpending 1986).

   Although the simple models presented above are useful for understanding the dynamic between gene flow and genetic drift, they are often *too* simple to apply to many organisms, especially humans. For example, the discussion above focused on F_{ST} being at equilibrium. Can we make this assumption when trying to interpret an F_{ST} value that has been estimated from genetic data? For example, if we analyze genetic marker or DNA marker data from a set of human populations and find an F_{ST} of 0.005, we can reasonably state that this is a low level of between-group variation, roughly on par with an estimated 50 migrants each generation.

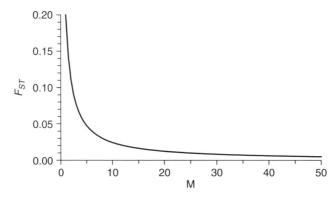

FIGURE 8.10 Expected equilibrium value of F_{ST} for different numbers of migrants per generation ($M = Nm$). Values for equilibrium F_{ST} were computed for values of M ranging from 1 to 50 using equation (8.7).

However, is the equilibrium assumption warranted? Perhaps we are sampling the population early in its existence and equilibrium has not been reached. How valid is the equilibrium assumption? The answer to that question depends on the nature of the populations being studied, as time, population size, and migration rate will all affect changes in F_{ST} over time.

Although it is useful to develop preliminary models based on a simple set of assumptions where population size and migration rate remain constant over time, there are many examples in human history where these demographic parameters have changed rapidly, particularly in modern times. The population of Ireland is a good example, where the population grew rapidly in the 1700s because of the introduction of the potato as a viable crop in the island's ecology, but then declined rapidly in the nineteenth century following the failure of the potato crop (the great Irish famine) (Connell 1950; Kennedy 1973). The types of demographic changes we see in human populations render assumptions of constancy questionable in many cases. Other models and methods, some of which are presented below, allow more flexibility in analyzing the interaction between gene flow and genetic drift. In addition to the theoretical aspects discussed here, more case studies from human populations will be discussed in Chapter 9.

C. Isolation by Distance

An assumption underlying the models presented earlier is that the number of migrants exchanged between subpopulations is the same. Although this can be a useful assumption for developing baseline models, it may not be very useful for applying to analyses of the real world where the numbers of migrants between populations can very significantly. One key influence on migration in many species, including humans, is geographic distance. Indeed, much of the genetic variation in our species can be explained by the geographic distance between populations (Relethford 2010) (see also Chapter 9).

Quite simply, you are more likely to choose a mate from close by than from farther away. Consider the fact that throughout most of human history and prehistory the amount one could travel in a day was often limited by how far/you could walk in a day. One was more likely to meet a potential mate close to home than from many miles away. This is still the case. Although we live in a world where rapid and distance transportation is of a magnitude far beyond that of our ancestors (you can cross the entire planet within a day!), we still tend for the most part to choose mates from close by. We tend to spend much more time living and working within a set geographic limit than crossing long distances. I illustrate this idea in my large introductory biological anthropology class by asking for a show of hands for the number of people whose parents were born in the same state. This accounts for a large proportion of the class. I also tend to get a lot of responses from those whose parents came from different but adjacent states. As I increase the distance to cover marriages in nonadjacent states or in other countries, I get fewer responses (but typically there are usually *some* long-distance marriages). Of course, my quick and dirty survey is not a valid scientific survey, nor is it intended to be. The purpose is to provide a personalized and participatory means of illustrating

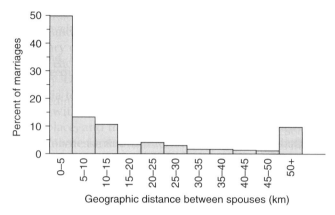

FIGURE 8.11 Marital distance distribution for north central Massachusetts, 1800–1849. Marital distance is the geographic distance (in kilometers) between the premarital residences of bride and groom, which gives us an idea of how far people move in search of mates. (*Source*: Author's unpublished data.)

the basic principle that geographic distance limits migration. A vast number of studies have shown this relationship across the range of human societies.

An example of the effect of geographic distance on mate choice is shown in Figure 8.11, which shows the geographic distance between the premarital residences of bride and groom in 3592 marriages that took place in four towns in north central Massachusetts from 1800 through 1849. These data are part of a study in the historical demography of population growth and population structure of the region (e.g., Relethford 1986, 1991). Half of the marriages took place between brides and grooms who lived less than 5 km from each other. The percentage of marriages declines quickly after that, with 13% of the marriages taking place between 5 and 10 km, 11% between 10 and 15 km, and 3% between 15 and 20 km. This graph shows the characteristic exponential decline in marriages (and therefore much of gene flow) over geographic distance. Note, however, that there is some long-distance migration, also characteristic of most human populations. Here, almost 10% of the marriages took place between spouses who were 50 km or more apart. Some of the marriages took place over very large distances—12 occurred over 500 km apart, including two at distances over 2000 km. This long-distance migration (and gene flow) helps keep populations from drifting apart too far even given the local limiting effect of geographic distance.

Isolation by distance models have been proposed in population genetics to examine the effect of geographic distance on genetic similarity between pairs of populations. Such models look at the balance between genetic drift, gene flow between local neighboring populations, and long-range gene flow on a measure of genetic similarity. Many different measures of genetic similarity have been proposed. Two measures that are used frequently in studies of human populations are the R matrix measure and Nei's genetic identity measure, described in more detail in Appendixes 8.2 and 8.3 for those interested in seeing different methods for computing genetic similarity.

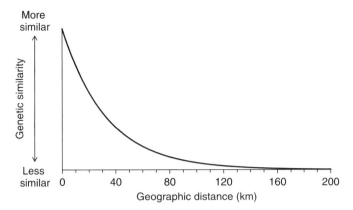

FIGURE 8.12 Isolation by distance. Isolation by distance models predict that genetic similarity between pairs of populations will decrease as the geographic distance between populations increases. This figure was obtained using the R matrix measure of genetic similarity (see Appendix 8.2) and the isolation by distance model $r_{ij} = (r_0 - r_\infty)e^{-bd} + r_\infty$ (Relethford 2004), where r_{ij} is the genetic similarity between populations i and j, and d is the geographic distance between populations i and j. The parameter r_0 is the genetic similarity at distance $d = 0$, r_∞ is the genetic similarity at very large values of d, and b reflects the amount of distance decay (Relethford 2004). For this graph, the parameters of the model are set to $r_0 = 0.05$, $r_\infty = -0.005$, and $b = 0.03$.

A number of different theoretical models of isolation by distance have been developed. They all share the basic idea that the genetic similarity between populations decreases as the geographic distance between populations increases. An example is shown in Figure 8.12. The greater the distance between two populations, the less gene flow between them is expected, and the level of genetic similarity will decline as a result of drift. Isolation by distance models typically incorporate some small level of long-range gene flow that moderates the decline in genetic similarity with distance. This causes the decrease in genetic similarity to level off after a certain distance. For example, in Figure 8.12, there is little change in genetic similarity after about 100 km. Genetic (and morphological) data have been found to fit the isolation by distance model in human populations around the world (Morton 1973; Jorde 1980; Relethford 2004). An example is shown in Figure 8.13, which looks at the relationship between genetic distance (based on red blood cell markers) and geographic distance between 10 Papago Indian populations in the American Southwest. Although there is scatter, the overall pattern of genetic similarity and geographic distance fits the isolation by distance model. Genetic similarity declines quickly within the first 50 km, and then continues to decrease at a slower rate at greater distances.

D. Migration Matrix Analysis

Another approach to examining the interaction of gene flow and genetic drift is to use observed rates of migration and population size to predict patterns of genetic variation that are expected when an equilibrium has been reached between these

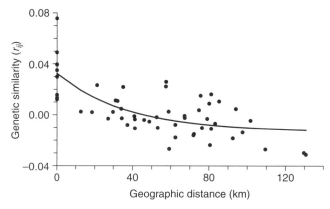

FIGURE 8.13 Isolation by distance among 10 Papago Indian populations based on red blood cell polymorphisms. The 10 populations are political districts on the Papago Indian reservation in Arizona. The black circles represent the genetic similarity and geographic distance between pairs of populations. The circles at zero geographic distance represent the genetic similarity within populations. The solid line is the fit of Relethford's (2004) isolation by distance model. Measures of genetic similarity were computed using the R matrix method (see Appendix 8.2) based on 22 alleles for nine red blood cell genetic markers reported by Workman et al. (1973). Geographic distances between districts were taken from Workman and Niswander (1970). The estimated parameters of Relethford's model [$r_{ij} = (r_0 - r_\infty)e^{-bd} + r_\infty$)] are $r_0 = 0.032, r_\infty = -0.013$, and $b = 0.027$.

two evolutionary forces. This approach is known as **migration matrix analysis**. In addition to providing some insight into the likely relative effects of gene flow and genetic drift, this predictive approach allows comparison with observed genetic data. For example, if we observe patterns of migration and population size in the present, we would be interested in seeing how well observed genetic variation fits that predicted by the migration matrix method. If observed and predicted patterns of genetic variation are not similar, then the observed demography (migration rates and population size) are likely to have changed in the past.

The heart of the method, the migration matrix, is defined as a matrix of migration probabilities where rows refer to population of origin and columns refer to population of residence. Each element of the matrix m_{ij} represents the probability that an individual in population j came from population i. A common source of such data is information on parent–offspring migration, where we would look at the birthplaces of parents and their offspring, such that m_{ij} represents the probability that a child born in population j had a parent born in population i. When data on parents and offspring are not available, we can also look at migration during an individual's life, such as classifying reproduction-aged adults by their population of origin and their population of residence. Here, m_{ij} refers to the probability that someone living in population j came from population i. In other cases, we can use data on migration at marriage to classify someone by according to where she or he lived before and after marriage. Here, m_{ij} would represent the probability that someone living in population j after marriage was living in population i before marriage. One advantage of working with human populations is that it is often easy to find out from human subjects where they (or their parents) lived.

Example 8.1 presents human migration data and computation of the migration matrix based on migration data from the Bedik tribe of eastern Senegal in Arica (Lanageny and Gomila 1973). The Bedik are a small ethnic group that lived apart from other Sengalese groups, and are culturally and linguistically distinct from many of their neighbors. As an example, while many neighboring tribes have converted to Islam, the Bedik have retained animistic beliefs (Jacquard 1974). Example 8.1 shows data on migration among six Bedik villages where each of 746 adults was classified according to village of origin (rows) and village of residence (columns). In order to construct the migration matrix showing the probability of an individual living in population j having been born in population i, each entry in the matrix is divided by its corresponding *column* total. For example, there were 30 individuals living in village 3 who came from village 4. The total number of people living in village 4 for which data were available was 194 (the column total). Therefore, the migration matrix entry $m_{43} = \frac{30}{194} = 0.1546$. This means that the probability of someone living in village 3 having come from village 4 is 0.1546. Also, note that a migration matrix is not necessarily symmetric; for example, the probability of living in village 4 and originating from village 3 ($m_{34} = 0.2581$) is greater than the probability of living in village 3 and originating from village 4 ($m_{43} = 0.1546$).

EXAMPLE 8.1   A Migration Matrix: The Bedik of Senegal.   The following data are for six villages among the Bedik of Senegal, Africa, where each adult has been classified according to village of origin and village of residence as reported by Langaney and Gomila (1973). Each element in the matrix below represents the number of individuals in population j (columns) that came from population i (rows):

Village of Origin	Village of Residence					
	1	2	3	4	5	6
1	47	10	1	0	6	0
2	1	105	2	0	1	3
3	4	5	144	40	23	3
4	1	1	30	100	1	12
5	5	16	12	2	38	1
6	1	3	5	13	1	109
Total	59	140	194	155	70	128

The elements of the migration matrix below (m_{ij}) represent the probability that a person in population j came from population i. These elements are computed by dividing each element in the matrix by the corresponding column totals. For example, 4 individuals in village 1 came from village 3. There are a total number of 59 individuals from village 1. The migration matrix value here is $m_{31} = \frac{3}{59} = 0.0678$:

Village of Origin	Village of Residence					
	1	2	3	4	5	6
1	0.7966	0.0714	0.0052	0.0000	0.0857	0.0000
2	0.0169	0.7500	0.0103	0.0000	0.0143	0.0234
3	0.0678	0.0357	0.7423	0.2581	0.3286	0.0234
4	0.0169	0.0071	0.1546	0.6452	0.0143	0.0938
5	0.0847	0.1143	0.0619	0.0129	0.5429	0.0078
6	0.0169	0.0214	0.0258	0.0839	0.0143	0.8516

The migration matrix provides information on local gene flow between the populations of interest. We also need to factor in the effect of long-range gene flow, which will act to help moderate the effect of drift. Here, *long-range gene flow* is typically defined as the proportion of individuals who came from outside the area used to construct the migration matrix. In the Bedik analysis, the 746 individuals that were used to construct the migration matrix all originated from one of the six villages in the study area. Langaney and Gomila (1973) also found 14 individuals living in the six villages that originally came from *outside* these six villages (as such, these 14 people do not appear in the migration matrix in Example 8.1 because the migration matrix only counts those whose residence *and* origin were in the six villages). Thus, we have a total number of $746 + 14 = 760$ people. The long-range migration rate is $\frac{14}{760} = 0.0184$. The final piece of information needed for a migration matrix analysis is the effective population sizes, which provides an estimate of the effect of genetic drift (more drift is expected with smaller effective population size). For the Bedik example, Langaney and Gomila (1973) computed effective population size for the six villages, which ranged from 45 to 145 with an average of 94.

Given data on local migration rates, long-range migration, and effective population sizes, migration matrix analysis is a mathematical method that allows prediction of genetic variation within and between populations over time. These predictions are performed in successive generations until the estimates of genetic variation reach an equilibrium between gene flow and genetic drift. Several different models of migration matrix analysis have been proposed using different measures of genetic variation and underlying assumptions (e.g., Bodmer and Cavalli-Sforza 1968; Smith 1969). The model used here, developed by Rogers and Harpending (1986), is a variation particularly well suited for application to human populations and for comparison with observed patterns of genetic variation. The actual mathematics are too involved for presentation here, but are presented in detail by Rogers and Harpending (1986).

Figure 8.14 shows the changes in F_{ST} predicted from Rogers and Harpending's migration matrix model applied to the observed patterns of migration and population size of the Bedik villages. The amount of between-group genetic variation (F_{ST}) increases rapidly to approach an equilibrium value within a few generations. After five generations there is little change in F_{ST}, with an equilibrium value of $F_{ST} = 0.0093$. The speed of convergence to equilibrium in a migration matrix analysis depends on the rates of local and long-range gene flow (Rogers and

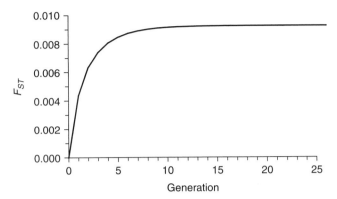

FIGURE 8.14   Predicted F_{ST} between Bedik villages based on a migration matrix analysis. The Rogers–Harpending method of migration matrix analysis was applied to the migration matrix presented in Example 8.1, and the estimate of long-range migration and effective population sizes given by Langaney and Gomila (1973). The analysis starts with an initial $F_{ST} = 0$, and the analysis is applied to successive generations until there is virtually no change in F_{ST} from one generation to the next. As shown here, F_{ST} increases rapidly at first as a result of genetic drift, and soon reaches a balance between drift and gene flow (local and long-range). For this example, the convergence to an equilibrium between drift and gene flow occurs within a few generations. As noted in the text, the speed of convergence is dependent on the intensity of local and long-range gene flow.

Harpending 1986; the higher the levels of gene flow, the faster the convergence of F_{ST}. This means that in cases of rapid convergence, there is a good chance that the observed migration–population size data are representative of the recent evolutionary past, and that the genetic predictions from the migration matrix analysis are more accurate. This is an important point in many human populations because demographic parameters such as population size and migration often change so quickly (consider, for example, the rapid growth in many human populations over the past century). In cases where convergence to equilibrium is slower, there is an increased chance that "current" demographic data do not reflect the past. For the Bedik example, I note that the predicted F_{ST} of 0.009 is close to the observed F_{ST} of 0.012 obtained from analysis of genetic markers (Rogers and Harpending 1986).

The most restrictive assumption of migration matrix analysis is that the method assumes that equilibrium has been reached. In cases where patterns of migration and population size have remained relatively stable and/or convergence to equilibrium is relatively fast (as with the Bedik), then the assumption of equilibrium is not too restrictive, and we would expect close correspondence between migration and genetics. Rogers and Harpending (1986) found that the correspondence between estimates of F_{ST} based on a migration matrix and genetic markets was close for cases where convergence to equilibrium was rapid, which were cases with higher rates of local and/or long-range migration. When convergence is slower, as will happen with lower rates of migration, then there is greater discrepancy between migration and genetic data. Therefore, the utility of the migration matrix approach is variable from one case to the next. Sometimes it is a useful approach, and sometimes it is not.

IV. ESTIMATING ADMIXTURE IN HUMAN POPULATIONS

The treatment of gene flow so far has concentrated on local gene flow for the most part, such as the genetic effects of gene flow from a continent to an island, or between villages located close to each other. Gene flow also takes place over much longer distances, as discussed briefly in the above sections on isolation by distance and migration matrix analysis. In the history of our species, in some cases there have been high levels of migration over very long distances, often between groups that typically were not part of the local exchange of genes through short-range gene flow. Many of these long-distance movements accompanied the geographic expansion of human beings over the past 500 years following the European age of exploration. As Europeans spread out from the Old World into the New World, they met previously unknown native populations. Sometimes this contact was violent, and sometimes it was peaceful, but in all cases it resulted in the mixing of gene pools from geographically distant populations. One example is gene flow from Europeans into the enslaved African-American population of colonial times. Another example is gene flow from Europe, primarily Spain, into Native American populations in Central America.

A. A Simple Admixture Model

Gene flow between previously separated (or partially separated) populations is known as **admixture**. The dynamics of admixture allow us to examine the genetic makeup of admixed populations and make inferences about the amount of ancestry from two or more sources. The simplest example of admixture is to consider two *parental* populations both contributing genes to a *hybrid* population. This model is presented in Figure 8.15, which shows some proportion of ancestry from population A and some from population B. If we label the proportion of

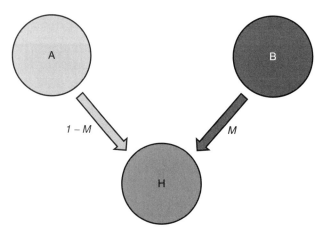

FIGURE 8.15 Admixture between two parental populations. The hybrid population (H) is formed by gene flow from two parental populations, A and B. The proportion of ancestry in population H from population B is symbolized by M, and the proportion of ancestry in population H from population A is therefore $1 - M$.

ancestry from population B as M, this means that the proportion of ancestry from population A must be $1 - M$ (because the total amount of ancestry must add up to 1). This means that for any given allele, we can describe the allele frequency of the hybrid (p_H) in terms of the allele frequencies of parental population A (p_A) and parental population B (p_B), each weighted by their contribution to ancestry:

$$p_H = (1 - M)p_A + Mp_B \qquad (8.9)$$

For example, if $p_A = 0.8$ and $p_B = 0.6$, and if the proportion of ancestry from population B were $M = 0.3$, then the allele frequency in the hybrid population would be

$$p_H = (1 - 0.3)(0.8) + (0.3)(0.6) = 0.74$$

Equation (8.9) is very useful for studying ancestry, because if we have estimates of the current allele frequencies for the hybrid populations and the two parental populations, we can estimate the ancestral contributions. We do this by expanding equation (8.9) to get

$$p_H = p_A - Mp_A + Mp_B$$

We then factor out M to obtain

$$p_H = p_A - M(p_A - p_B)$$

and then use some simple algebra to get

$$M = \frac{p_A - p_H}{p_A - p_B} \qquad (8.10)$$

For example, if we have estimates of $p_H = 0.7$ in the hybrid population and $p_A = 0.9$ and $p_B = 0.4$ for the parental populations, we would estimate the amount of ancestry in H from population B as

$$M = \frac{0.9 - 0.7}{0.9 - 0.4} = 0.4$$

Thus, 40% of the ancestry in the hybrid population came from population B, which means that the remainder (60%) came from population A. In any actual study, we would not want to rely on an estimate from a single allele, but instead would average these estimates over a large number of alleles, or use an admixture estimation method that is designed for multiple alleles and loci.

Keep in mind that M represents the accumulated admixture over past generations and is not the amount of admixture that occurs in a *single* generation. These two views of ancestry are easily confused. The amount of admixture per generation, which we will denote by lowercase m, represents how much ancestry comes from population B in a single generation. Accumulated admixture M represents how much ancestry from population B has *accumulated* over past generations. The difference between these two quantities is analogous to the amount of interest you would pay on average every year on a loan versus the total amount of interest paid over the lifetime of a loan. Per generation admixture m and accumulated admixture M are related as

$$M = 1 - (1 - m)^t \qquad (8.11)$$

where t is the number of generations that have passed since admixture began (and assuming that m is constant from one generation to the next). The derivation of equation (8.11) is given in Appendix 8.4. As an example, if the per generation admixture rate from population B were $m = 0.01$ and remained constant for $t = 20$ generations, the accumulated admixture from population B would be

$$1 - (1 - 0.01)^{20} = 0.18$$

(18%). Note that our estimates of admixture from genetic data are of accumulated admixture M. However, if we have an idea of the number of generations that have passed (say, from historical data), we can estimate the average per generation admixture by rewriting equation (8.11) as

$$m = 1 - e^{\ln(1-M)/t}$$

where ln is the natural logarithm and e is the base of natural logarithms ($= 2.71828$). For example, if we had an estimate of $M = 0.3$ from genetic data and knew that admixture had occurred over the previous 10 generations, our estimate of per generation admixture would be $m = 0.035$. Like interest on a loan, total ancestry can add up generation after generation (assuming that the rate of admixture remains constant).

B. Assumptions of Admixture Estimation

As with any population-genetic model, the admixture estimation model makes several assumptions (Reed 1969). The first assumption is that we have identified the parental populations correctly. Calculation of the number and identity of parental populations can be facilitated by cultural and historical information. For example, history shows us that the major source of initial contact between native populations and Europeans in Mexico following the European age of exploration was Spain. It makes sense, therefore, to use data on Spanish allele frequencies to represent the European source and not France or England. Likewise, if we are investigating admixture in African-Americans, we need to choose African allele frequencies from the geographic regions from which native peoples were captured and enslaved, which would be west central Africa.

The second assumption listed by Reed (1969) is that no allele frequency change has occurred in the parental population from the time of initial admixture until the present. This assumption is necessary because we use the allele frequencies that we observe in the present as a proxy for the allele frequencies in the past. If the parental allele frequencies have changed, then our admixture estimates will be off. We are assuming a stable equilibrium in the parental populations that has not been affected significantly by mutation, drift, selection, or gene flow from another population. Our third assumption is that admixture is the *only* cause of allele frequency change in the hybrid population. In other words, we assume that there has been no significant mutation, selection, and/or drift in the hybrid population so that the present-day allele frequency has been shaped only by the admixture process. The fourth assumption given by Reed is that all of our present-day allele frequencies are unbiased estimates and have not been affected significantly by sampling error (which would happen if we estimated allele frequencies from small samples).

Although critical, these assumptions are not insurmountable. As noted above, cultural and historical context can provide us with needed clues about the geographic and ethnic identity of ancestral populations, and competing models (e.g., whether there is a third parental population) can be tested to see which offers the best fit to the data. Sampling error can be handled to some extent by choosing large samples when possible and estimation over as many different alleles as possible. Variation in admixture estimates due to sampling error and genetic drift can now be estimated using more advanced methods of admixture analysis (Long 1991).

Another useful approach is to use genetic markers that are widely different in the parental populations, known as **ancestry-informative markers** (AIMs). If the allele frequencies in the parental populations are similar to each other to begin with, then any source of variation other than admixture (such as drift) can alter one or both parental allele frequencies enough to make accurate estimation of admixture difficult. By using alleles with large differences between parental populations, the researcher can minimize any changes in allele frequencies due to other sources of variation. The recent and rapid development of huge numbers of DNA markers, such as single-nucleotide polymorphisms (SNPs), has provided numerous AIMs for detecting ancestry (e.g., Rosenberg et al. 2003a; Seldin and Price 2008).

C. Extensions for Admixture Analysis

Although useful, estimation of admixture from equation (8.10) is somewhat limited. For one thing, the simple admixture model has only two parental populations, and there have been cases in human history of three or more major sources of ancestry. A number of admixture methods have been developed that allow for three (or more) parental populations. An example is Devor and Crawford's (1980) analysis of admixture in Tlaxcaltecan Indian populations in Mexico. Although the most significant sources of ancestry were from Native Americans and Spaniards, several populations also showed significant ancestry from western Africa, presumably introduced through enslaved Africans who had been brought in to work on local mines.

A number of different methods have been developed that allow estimation of admixture proportions using data from multiple alleles and loci. One of the most useful has been the method outlined by Long (1991) that factors in the effect of sampling bias and variation introduced by genetic drift. His method also provides a statistical test of variation across different loci, which can be useful in detecting natural selection. Whereas gene flow and genetic drift affect all loci, natural selection at one locus may not have any effect on other loci.

Thus far, we have been looking at admixture estimation and proportions of ancestry at the level of the local population. We need to keep in mind that these proportions are averages, and not everyone in the population may have the same levels of ancestry. Imagine, for example, a sample of five individuals in a hybrid population formed from parental populations A and B, and we estimate that the average amount of ancestry in the sample from B was $M = 0.2$, or 20%. Does this mean that *each* of the five people have 20% ancestry from B? It is possible, but another possibility might be two people with 0% ancestry from B, two people with 30% ancestry from B, and one person with 40% ancestry from B. The average here would also be 20%: $(0 + 0 + 30 + 30 + 40)/5 = 20$. A third case might be three people with 0% ancestry from B and two people with 50% ancestry from B, which also gives an average of 20%: $(0 + 0 + 0 + 50 + 50)/5 = 20$. The point here is that the average genetic ancestry of a population does not necessarily apply equally to all individuals within the population.

Several methods have been developed to allow estimation of *individual* ancestry (e.g., Hanis et al. 1986). One of these has been devised by Pritchard and colleagues and implemented in a computer program known as STRUCTURE (Pritchard et al. 2000; Falush et al. 2003). Here, individuals are assigned to one or more populations using a probability model based on their genotype. If they are admixed, they can be assigned to two or more populations, allowing an estimation of their ancestry. Examples of admixture studies looking at both group averages and individual ancestry are presented in the next chapter.

V. SUMMARY

Gene flow can affect variation both within and between populations. Gene flow can introduce new alleles into a population from elsewhere, thus increasing genetic diversity within a population. Gene flow acts to reduce variation between populations as it makes the allele frequencies in different populations more similar over time. One of the simplest models of gene flow, the island model, shows how the genetic makeup of a population can change under one-way gene flow, such that the recipient population (the "island") will eventually have the allele frequencies in the source population. For most studies of human populations, two-way gene flow is more applicable. Here, allele frequencies in all populations change as a result of sharing genes, and allele frequency differences become smaller over time.

Gene flow and genetic drift operate in opposition to each other. Genetic drift acts to make populations more different, and gene flow acts to make them more similar. Over time, an equilibrium is reached in terms of between-group variation (F_{ST}). A number of models and methods have been developed to investigate the balance between gene flow and genetic drift. The isolation by distance model is based on the idea that geographic distance limits gene flow, such that much of the gene flow into a population comes from nearby. As a consequence, genetic similarity between populations declines as the geographic distance between them increases. Migration matrix analysis is another way of studying gene flow in human populations, where observed data on migration are used to derive the probability that an individual living in one population came from another population. Combined with data on long-range migration and effective population sizes, the migration matrix allows prediction of the expected patterns of genetic variation hypothesized from an interaction of gene flow and genetic drift given observed data.

Another type of analysis of gene flow is the study of admixture, the level of gene flow into a hybrid population formed from the contact of two or more previously distinct populations. Admixture analysis allows estimation of the ancestry of samples and individuals, such as the amount of European ancestry in the gene pool of African-Americans.

APPENDIX 8.1 CHANGES IN ALLELE FREQUENCY OVER TIME IN AN ISLAND MODEL

The simplest formulation of the island model considers one-way gene flow. Here, an island experiences gene flow from the mainland, but not the reverse. As shown in the main (chapter) text, let p_0 be the initial allele frequency of the island. Let P be the allele frequency on the mainland, assumed to remain constant over time (because in a model of one-way gene flow, the mainland affects the island, but not the reverse). Each generation, let the island receive a proportion m of its alleles from the mainland, and a proportion $(1 - m)$ of its alleles from itself (see Figure 8.1). Using a simple model of mixing, the allele frequency of the island in the next generation ($t = 1$) is equal to

$$p_1 = (1 - m)p_0 + mP \tag{A8.1}$$

This is an iterative equation of the form

$$x_t = x_{t-1}(a + 1) + b \tag{A8.2}$$

seen in equation (A4.1). As shown in Appendix 4.1, an iterative equation of this form can be expressed as a function of time t as

$$x_t = (a + 1)^t \left(x_0 + \frac{b}{a} \right) - \frac{b}{a} \tag{A8.3}$$

Comparing equations (A8.1) and (A8.2) shows that $b = mP$ and $(a + 1) = (1 - m)$ which, in turn, means that $a = -m$. Substituting these values into equation (A8.3) and letting $x_0 = p_0$ and $x_t = p_t$ gives

$$p_t = (1 - m)^t \left(p_0 + \frac{mP}{-m} \right) - \frac{mP}{-m}$$

This simplifies to

$$p_t = (1 - m)^t(p_0 - P) + P$$

which is equation (8.1). Note that as t increases, the quantity $(1 - m)^t$ gets smaller and smaller, approaching zero. As this happens, the above equation shows that

p_t approaches P, which is the allele frequency of the mainland. In other words, the genetic makeup of the island is eventually replaced by immigration from the mainland.

APPENDIX 8.2 GENETIC SIMILARITY: THE R MATRIX

An R matrix provides a measure of genetic similarity between any pair of populations in an analysis (including a population compared with itself). Let g be the number of populations. The R matrix will have g rows and g columns. Each element in the R matrix (r_{ij}) represents the genetic similarity between populations i and j (referenced using the subscripts). For any given allele, the elements of the R matrix are computed as

$$r_{ij} = \frac{(p_i - \bar{p})(p_j - \bar{p})}{\bar{p}(1 - \bar{p})} \tag{A8.4}$$

where p_i is the allele frequency in population i, p_j is the allele frequency in population j, and $\bar{p}$ is the mean allele frequency computed using *all* of the populations in the analysis (not just i and j). The numerator expresses how i and j covary in their difference from the mean, and the denominator is a way of standardizing (Harpending and Jenkins 1973). When $i = j$, the value of r_{ij} refers to genetic similarity within a group, and when $i \neq j$, the value of r_{ij} indicates genetic similarity between groups. The R matrix is symmetric, which means that $r_{ij} = r_{ji}$.

Equation (A8.4) applies for any single allele. In any actual analysis, the R matrix needs to be based on as many alleles and loci as possible, and the R matrix is obtained by averaging equation (A8.4) over all alleles. Other computational details are important in some cases. If population size is known to vary across the populations, then the mean allele frequency $\bar{p}$ should be computed using a weighted average. There are also methods of adjusting for bias due to sample size [see Harpending and Jenkins (1973) and Workman et al. (1973) for more details on R matrix computation]. In addition to use with allele/haplotype frequencies, methods have been developed that allow estimation of R matrices from migration data (Rogers and Harpending 1986), surname frequencies (Relethford 1988), and quantitative traits (Williams-Blangero and Blangero 1989; Relethford and Blangero 1990; Relethford et al. 1997). These extensions make R matrix analysis very useful for comparing patterns of genetic similarity from different types of data.

As R matrices express genetic similarity in terms of deviations from the mean allele frequencies, the average value of all elements of an R matrix (weighted by population size if needed) is equal to zero. Values of r_{ij} that are *greater* than zero indicate a pair of populations that are *more* similar to each other than on average. Values of r_{ij} that are *less* than zero indicate a pair of populations that are *less* similar to each other than on average.

The diagonal elements of the R matrix ($i = j$) are a measure of genetic similarity *within* populations, and will always be positive (and generally the largest values in the R matrix) because individuals are more similar to each other *within* than *between* populations. A handy feature of R matrices is that the average value of the diagonal elements (weighted by population size if needed) is an estimate of F_{ST}.

The properties of an R matrix are best illustrated with an example. Imagine three populations with the following allele frequencies: $p_1 = 0.70$, $p_2 = 0.72$, and $p_3 = 0.76$. We will assume the three populations are the same size, so we can compute the mean allele frequency as a simple average: $\bar{p} = (0.70 + 0.72 + 0.76)/3 = 0.7267$. We can now compute the elements of the R matrix using equation (A8.4):

$$r_{11} = \frac{(0.70 - 0.7267)(0.70 - 0.7267)}{0.7267(1 - 0.7267)} = 0.0036$$

$$r_{12} = \frac{(0.70 - 0.7267)(0.72 - 0.7267)}{0.7267(1 - 0.7267)} = 0.0009$$

$$r_{13} = \frac{(0.70 - 0.7267)(0.76 - 0.7267)}{0.7267(1 - 0.7267)} = -0.0045$$

$$r_{22} = \frac{(0.72 - 0.7267)(0.72 - 0.7267)}{0.7267(1 - 0.7267)} = 0.0002$$

$$r_{23} = \frac{(0.72 - 0.7267)(0.76 - 0.7267)}{0.7267(1 - 0.7267)} = -0.0011$$

$$r_{33} = \frac{(0.76 - 0.7267)(0.76 - 0.7267)}{0.7267(1 - 0.7267)} = 0.0056$$

Because the matrix is symmetric, we can fill in the remaining values (e.g., $r_{21} = r_{12}$). The entire R matrix looks like this:

	1	2	3
1	0.0036	0.0009	−0.0045
2	0.0009	0.0002	−0.0011
3	−0.0045	−0.0011	0.0056

As is the case with R matrices, the average value of any row or column, as well as the entire matrix, is zero. The average of the diagonal elements gives an estimate of F_{ST} of $(0.0036 + 0.0002 + 0.0056)/3 = 0.0031$. Several other observations apply here:

1. Population 2 has the lowest r_{ii} value (0.0002), which means that it has the allele frequency closest to the mean.
2. Population 3 has the highest r_{ii} value (0.0056), which means that it has the allele frequency most different from the mean.
3. When we look at all of the off-diagonal values (which are comparisons *between* populations), we see that the two lowest values (indicating *less* genetic similarity) are associated with population 3, and the highest r_{ij} value is between populations 1 and 2 (0.0009). These observations mean that populations 1 and 2 are more similar genetically than is population 3.

It may occur to you that all of these observations could more easily been made just looking at the actual allele frequencies and that there was no need to go through the entire analysis! This is true, but only because we were looking at a single allele. If we had more alleles, then the situation could get complicated very quickly, particularly if there is variation in results from one allele to the next. R matrix analysis allows inferences to be made using all alleles in any analysis, summarizing the *average* patterns of genetic similarity.

It is sometimes useful to consider the amount of genetic *dissimilarity* between populations, referred to as a *genetic distance*. For any given allele, one common measure of genetic distance between populations i and j can be computed as

$$D_{ij}^2 = \frac{(p_i - p_j)^2}{\bar{p}(1 - \bar{p})}$$

which is the standardized squared difference in allele frequencies. An overall squared distance can be averaged over all alleles. The squared distances can also be derived directly from the R matrix as

$$D_{ij}^2 = r_{ii} + r_{jj} - 2r_{ij}$$

Note that when $i = j$, the squared distance is zero (which makes sense, as the genetic distance of a population to itself must be zero). For some applications, it is useful to look at the squared distance (D^2), and in some cases it is useful to look at the unsquared distance $D = \sqrt{D^2}$.

APPENDIX 8.3 GENETIC SIMILARITY: NEI'S GENETIC IDENTITY

A widely used measure of genetic similarity within and between populations has been proposed by geneticist Masatoshi Nei (1972, 1987). For any given locus and allele, let x and y be the allele frequencies in populations X and Y. For any given locus, a measure of genetic similarity is the probability that a random allele from X will be identical to a random allele from Y, which is

$$j_{XY} = \sum xy$$

where summation is over all alleles at that locus. This measure is then averaged over all loci to calculate the overall quantity J_{XY} (the lowercase j is used to refer to any given locus, and the uppercase J refers to the average over all loci). When looking at variation within population X or population Y, the relevant quantities are

$$j_X = \sum x^2$$

and

$$j_Y = \sum y^2$$

which are then averaged over all loci to get J_X and J_Y.

Nei's measure of genetic identity I (also known as the *normalized identity of genes*) between populations X and Y is then computed as

$$I_{xy} = \frac{J_{xy}}{\sqrt{J_x J_Y}}$$

Nei shows that a genetic distance between populations X and Y can be computed as

$$D_{xy} = -\ln(J_{xy})$$

where ln denotes the natural logarithm.

APPENDIX 8.4 RELATIONSHIP BETWEEN PER GENERATION ADMIXTURE (M) AND ACCUMULATED ADMIXTURE (M)

Equation (8.9) was derived considering the relationship between the allele frequencies of the hybrid and parental populations in terms of accumulated admixture (M) that occurred after some number of generations. If we look at the problem on a generation-by-generation basis, we can model admixture using the island model. Picture a founding population from parental population A with allele frequency p_A that forms the beginning of the hybrid population. Now, with each generation, a proportion of the alleles in the hybrid population is from population B with allele frequency p_B at rate m. Assuming that the allele frequency for population B does not change over time, we can calculate the allele frequency in the hybrid population in any generation (p_{H_t}) as a function of the hybrid allele frequency in the previous generation ($p_{H_{t-1}}$), the per generation admixture rate (m), and the allele frequency in parental population B (p_B) as

$$p_{H_t} = (1 - m)p_{H_{t-1}} + mp_B$$

Because this is an iterative equation, we can use the methods outlined in Appendix 4.1. In the case of admixture, this equation is the same form as equation (A8.1) in Appendix 8.1, and the solution is

$$p_{H_t} = (1 - m)^t (p_{H_0} - p_B) + p_B$$

which expresses the allele frequency of the hybrid after t generations of admixture in terms of the initial allele frequency of the hybrid p_{H_0}. Now, remember that this initial founding population came from parental population A, which means that the initial allele frequency in the hybrid population is equal to the allele frequency in parental population A, or $p_{H_0} = p_A$. We can now rewrite the above equation completely in terms of the assumed unchanging allele frequencies of the parental populations, A and B, as

$$p_{H_t} = (1 - m)^t (p_A - p_B) + p_B$$

which, in turn, we can rewrite as

$$(1 - m)^t = \frac{p_{H_t} - p_B}{p_A - p_B}$$

We are now ready to rewrite this equation in terms of the observed allele frequency of in the hybrid population after t generations of admixture (which we have denoted p_H), which gives

$$(1 - m)^t = \frac{P_H - P_B}{P_A - P_B} \qquad (A8.5)$$

Recall from equation (8.10) in the main chapter text that

$$M = \frac{p_A - p_H}{p_A - p_B}$$

If we subtract this value from 1, we get

$$1 - M = 1 - \left(\frac{p_A - p_H}{p_A - p_B}\right)$$

$$1 - M = \left(\frac{p_A - p_B}{p_A - p_B}\right) - \left(\frac{p_A - p_H}{p_A - p_B}\right)$$

$$= \frac{(p_A - p_B) - (p_A - p_B)}{p_A - p_B}$$

$$= \frac{p_H - p_B}{p_A - p_B}$$

Note that the right side of this equation is the same as the right side of equation (A8.5), which means that

$$(1 - m)^t = (1 - M)$$

which can be rewritten as

$$M = 1 - (1 - m)^t$$

which is equation (8.11) in the main text.

CHAPTER 9

HUMAN POPULATION STRUCTURE AND HISTORY

I have always been interested in the question of origins and history, ranging from my own family history to the evolution of our entire species. I believe it is common for people to think about their past and how they fit into history, be it family history, national/cultural history, or global history. We might think of history in many terms, including economics, politics, and social change. I suspect that not many people think about history in terms of the mathematical relationships of microevolution that have been presented in this book. Our earliest training in history in schools tends to focus on the past few thousand years in terms of written history, yet to an anthropologist, history is much deeper than the timelines covered in written records.

From the perspective of population genetics, history can be studied by looking at the record of genetic diversity. Events in the past have in many cases left their signature in our DNA, and we are often able to learn a lot about the structure and history of our populations by interpreting genetic diversity in terms of the evolutionary forces. For example, cultural and linguistic differences between human groups, often studied in terms of the social and economic correlates of culture contact, also affect rates of gene flow. Changes in population size due to ecological and demographic changes will, in turn, affect levels of genetic drift.

This final chapter examines more closely the relationship of human population structure and population history to patterns of genetic variation within and between populations. Some aspects of history have already been covered in the chapters on natural selection in terms of the history of specific genes or DNA sequences. Here, the focus is instead on patterns of neutral variation as assessed over many genes and haplotypes (and, in some cases, over the entire genome). Our interest here is on how mutation, genetic drift, and gene flow reflect the past structure and history of human populations. It is in these studies that the effect of the anthropologist's concern with how cultural behavior affects genetic variation is clearest. This chapter examines population structure and history by focusing on a set of case studies.

Human Population Genetics, First Edition. John H. Relethford.
© 2012 Wiley-Blackwell. Published 2012 by John Wiley & Sons, Inc.

In broad terms, **population structure** refers to the impact of mate choice and population composition on patterns of genetic variation within a population or patterns of genetic differences between subdivisions of a population. **Population history** is concerned more with the effect of historical events, such as population movements and culture contact, on patterns of genetic variation between populations. These two approaches to neutral genetic variation in humans overlap considerably (Harpending and Jenkins 1973), but a distinction between them is useful for organizing the case studies presented in this chapter.

I. CASE STUDIES OF HUMAN POPULATION STRUCTURE

The three case studies chosen here (out of hundreds, if not thousands, of possibilities) show how culture, environment, and geography have affected the genetic structure of local populations in three parts of the world—New Guinea, northwestern Europe, and South America.

A. The Gainj and Kalam of New Guinea

A number of studies have investigated the patterns of migration and genetic diversity in the Gainj- and Kalam-speaking Melanesian populations that live in the northern part of the central highlands of Papua New Guinea (e.g., Wood et al. 1985; Long 1986; Long et al. 1987; Smouse and Long 1992). The Gainj and Kalam are neighboring groups that speak different languages, but live in a similar environment and are both tribal horticulturists. The Gainj and Kalam live in small groups known as *parishes* that range in size from 20 to 200 individuals. Migration and genetic data have been collected for 18 parishes. Geographically, these parishes are arranged in a more or less straight line within river valleys, with predominantly Gainj speakers at one end, predominately Kalam speakers at the other end, and several parishes that have a mixture of languages in between. People travel from one parish to the next along a narrow dirt trail. All parishes are within 40 km walking distance of each other (Wood et al. 1985).

Studies of Gainj and Kalam population structures have concentrated on how the overall magnitude of migration, how migration is affected by geographic distance and language differences, and whether these factors have had an impact on genetic variation. In terms of migration, Wood et al. (1985) found quite a difference in rates of parent–offspring migration between males and females. Among males, 84% were born in the same parish as their father, whereas among females, only 33% were born in the same parish as their mother. This sex difference is due to patrilocal postmarital residence, which means a married couple will tend to take up residence in the husband's parish of birth, so that women are more likely to move. Even so, movement among both males and females tends to be highly restricted by geography. Among the actual migrants, the average distance from premarital to postmarital residence was about 8 km for males and 6 km for females. Far fewer migrations took place over long distances.

Wood et al. (1985) also looked at migration using a model that factored in geographic distance, linguistic differences (in terms of the percentage of Gainj and Kalam speakers in a parish), and population size. All of these factors had some effect on the rate of migrations between pairs of parishes. For both males and females, geographic distance and language differences had an effect, although in different ways. For males, language differences were more important in restricting migration than geographic distances, whereas for females the two factors had an equal influence. Differences in population size influenced migration rates in both sexes, with a tendency for individuals to move from larger populations to smaller populations more often than the reverse. This pattern might be a form of population size regulation in New Guinea, although this is not clear, as this effect of population size has been seen in a variety of human populations of different size and ecology (Relethford 1992).

Looking at genetic differences, Long (1986) found a value of $F_{ST} = 0.0224$ based on genetic marker data from the Gainj and Kalam parishes. This is a moderate level of genetic variation between groups given the fact that the parishes are not that far apart geographically and there is a high rate of migration outside one's birth parish. Closer analysis suggested that the underlying population structure is more complex, where each parish is actually made up of a number of smaller breeding populations.

Long et al. (1987) and Smouse and Long (1992) focussed on the relationship of genetic distances between populations based on genetic markers and geography, language, and population size. They found that the smaller parishes were more genetically divergent from other parishes, which is expected given genetic drift. Geographic distance and language differences had little effect on genetic distances between populations. Although geography and language did affect levels of migration, these effects were not seen in the actual genetic distances. Instead, the only significant impact on the actual genetic distances was population size—smaller populations showed greater genetic drift. Smouse and Long (1992) argue that the relatively high rate of overall migration effectively "erases" previous population structure and the observed genetic differences reflect only the most recent effects of genetic drift. When we observe genetic differences between populations, we see a record of past demographic history, but sometimes, as is the case here, that history is relatively shallow.

An important thing to remember about studies of human population structure is that they are not necessarily representative of an entire geographic region of culture. Results can vary from one study to another depending on specifics of geography and culture. On one hand, the relatively low correlation of geography and language with genetics among the Gainj and Kalam have been found in other populations of highland New Guinea, so it could be argued to be typical of that region (Long et al. 1987). On the other hand, studies of villages on Bougainville Island, elsewhere in Melanesia, showed higher correlations of geography and language and genetics (Friedlaender 1975).

B. The Åland Islands

Another example of how migration and genetic data can be used to study human population structure comes from across the world in the Åland Islands, which are located between Sweden and Finland in northern Europe generations (Jorde et al. 1982; Mielke et al. 1976, 1982). The Åland Islands consist of over 6000 islands and skerries (rocky islands that are too small for inhabitation). Several centuries of migration data have been collected from archival sources that allow investigation of changes in expected population structure over a number of generations. In addition, a large amount of genetic marker data were collected between 1958 and 1970 representing three recent generations.

The overall picture seen in the Åland Island studies is one of change associated with the breakdown of isolation following cultural changes. These trends are clear on examination of migration data based on marriage records from 15 parishes. For example, Relethford and Mielke (1994) found that the average level of exogamy among the 15 parishes remained relatively stable from the mideighteenth century (9.0%) to the end of the nineteenth century (7.1%). During the most recent time period (1900–1949), however, the average exogamy rate increased dramatically to 14.7%. This change is likely related to a number of demographic and economic changes affecting availability of potential mates as well as improved transportation opportunities in the early twentieth century, such as the introduction of steamboat routes and an increase in privately owned motorboats.

Analysis of the marriage records from 1750 to 1949 showed that both geographic distance and population size had an effect on levels of migration between Åland Island parishes. As is typically the case for human populations (and many other species), migration rates declined with geographic distance; more individuals choose mates closer to home than from more distant parishes. Although geographic distance affected migration rates for all time periods studied, the *rate* of decline in migration with distance changed over time, such that more migration took place over longer distances in the twentieth century than in earlier times (Mielke et al. 1994). This change also reflects the changes resulting in greater mobility that took place in the early twentieth century. As people became more mobile and travelled farther, the exponential decline of migration with geographic distance became less acute, and the slope of distance decay leveled off.

The availability of data on martial migration patterns allowed Mielke et al. (1976, 1982) to use a migration matrix approach for predicting how gene flow and genetic drift might have changed over time. As expected, the predicted values of genetic similarity within and between populations fit the isolation by distance model, as expected given the effect of geographic distance on migration. The increase in migration over time, as well as an increase in population size, suggests that genetic drift declined and gene flow increased, resulting in reduced levels of genetic similarity between populations. Similar results were found using data on parent-offspring migration (Jorde et al. 1982).

Of course, the migration matrix approach can only tell us what we will expect to see at equilibrium between gene flow and genetic drift based on observed patterns of migration and population size. As noted in Chapter 8, the *actual* patterns of genetic variation might be different. For the Åland Islands, genetic marker data were available to examine how well the predictions from migration matrix analysis fit reality. Jorde et al. (1982) took data collected between 1958 and 1970 and collated it by each person's year of birth, allowing them to sample three generations: (1) those born before 1900, (2) those born between 1900 and 1929, and (3) those born after 1929. They also examined genetic variation separately for the parishes on the main island as well as between parishes located in the more isolated outer islands. This strategy allowed Jorde and colleagues to look at changes in genetic variation over both time and space.

Using genetic markers, Jorde and colleagues found that F_{ST} was greater in the outer islands than on the main island, as expected given their greater isolation and smaller effective population size. However, there was little change in F_{ST} over time, unlike the expectations from migration matrix analysis. The same difference between migration and genetic data was found using isolation by distance analysis; the migration data showed a clear reduction in isolation over time, but this trend was less clear in the genetic data. The assumption of equilibrium used in migration matrix analysis may not appropriate, given the rapid demographic changes that took place in the Åland Islands during the twentieth century. Even so, further analysis showed that genetic variation was related to both population size and migration on attempts to separate the two effects (important because migration matrix analysis looks at both drift and gene flow at the same time). Here, Jorde et al. (1982) looked at the correlations between genetic distance and distances on the basis of a simple expectation of drift (based on population size differences) and a simple expectation of gene flow (based on migration rate differences). They found that drift had a greater effect in the earliest time period, but declined over time, and gene flow had a greater impact on genetic variation in the most recent time period. Again, this change is consistent with the observations of increasing migration rates (increasing the effect of gene flow) and increasing population size (reducing the effect of genetic drift).

The Åland Island studies comparing genetic and demographic data are important because they show us that some population genetic models, such as the migration matrix and isolation by distance models, may sometimes make assumptions that do not match up with the reality of rapid demographic and cultural change in human populations, particularly in more recent times. Still, careful examination of results can show us that genetic variation is shaped by gene flow and genetic drift, and the relative impact of these forces can change as well.

C. Altitude and Population Structure in Jujay, Argentina

A major (perhaps *the* major) influence on human population structure is geographic distance (e.g., Jorde 1980). As shown in the examples above and in Chapter 8, this makes sense, given the limiting effect that geographic distance has on migration. While this simple model is appealing, in many cases it may be too simple, as it ignores other influences on gene flow and population structure. The developing field of landscape genetics considers other ecological influences on gene flow and genetic variation, such as rivers and mountains (Storfer et al. 2007).

An example of an environmental influence on population structure is given here from a study of altitude and population structure in the Province of Jujay in northwestern Argentina (Gómez-Pérez et al. 2011). This study examined variation in a type of DNA marker known as an *Alu insertion*, which is a short section of DNA that replicates and moves to different locations on other chromosomes. Alleles are defined by the presence or absence of different insertion markers. *Alu* insertions are useful in tracking ancestry because the ancestral condition is the absence of an insertion, and any two individuals sharing the same *Alu* insertion marker would have inherited it from a common ancestor. Gómez-Pérez and colleagues looked at eight different *Alu* insertion markers in five regions of Jujay, sampling a wide range of altitude ranging from about 1200 meters (3937 ft) to over 3300 m (10,827 ft). The purpose of this study was to examine the impact of altitude on genetic diversity and genetic distance.

Gómez-Pérez et al. (2011) found that heterozygosity was lower in the high-altitude regions than in those at lower altitudes. Population density is lower in the high-altitude populations, contributing to increased genetic drift. In addition, the high-altitude populations have lower levels of exogamy, which means that there is less gene flow into the populations. The combined effect of increased genetic drift and decreased gene flow has resulted in lower levels of heterozygosity. In addition, the sources of gene flow also varied by altitude. The genetics of Jujay reflects a mix of Native American, European, and African ancestors, where European settlers and enslaved Africans entered the gene pool between the sixteenth and nineteenth centuries. Using the *Alu* insertion markers, Gómez-Pérez and colleagues found that Native American ancestry increased with altitude. Populations at the highest altitude had 100% Native American ancestry, whereas the lower-altitude populations showed less Native American ancestry but had significant European and African ancestry. Isolation by altitude appears to have affected levels of gene flow that entered the native populations over historic times. Immigrants tended to settle at lower altitudes, leading to more admixture.

Gómez-Pérez et al. (2011) also looked at the influence of altitude on genetic distances between the populations. They found a moderate and significant correlation between genetic distance and altitude difference; the farther apart two groups in terms of altitude, the greater the genetic distance between them. They also found a significant correlation with geographic distance, which was much stronger than the correlation with altitude. They conclude that the spatial distribution of *Alu* insertion markers in Jujay was best explained in terms of the isolation by distance model, further affected by differences in altitude. As with the studies of the Gainj/Kalam and the Åland Islands, the Jujay study shows the importance of using information on unique environmental, demographic, and cultural aspects of population structure to understand genetic variation.

II. THE ORIGIN OF MODERN HUMANS

The remaining case studies in this chapter focus on an examination of how the particular history of a population is reflected in patterns of genetic variation. A common aspect of the case studies is the use of genetic data to provide us with an understanding of population origins. Several different levels of analysis will be presented, starting with the most inclusive—the genetic history of our entire species. Additional case studies look at the origins of different populations around the world.

When most people think about the long-term evolution of our species from earlier human ancestors, they typically think of the fossil record. Although the fossil and archaeological records are an important source of information on human evolution, the last several decades have seen increasing attention to the role that genetic variation can play in understanding human origins. When we study the genetic variation within and between human populations in the world today, we can detect the genetic signature of many past events in our evolutionary history. Further, advances in technology have now allowed direct estimates of past genetic variation through the analysis of ancient DNA extracted from fossil remains tens of thousands of years old. The use of genetics in analyzing evolutionary events was described briefly in Chapter 4 in terms of the split between the African ape and human lines. Here, the topic of how genetics informs us about the origin of modern humans will be reviewed briefly.

A brief review of the major stages of human evolution is needed to focus on the origin of modern humans. Bipedal ancestors arose in Africa approximately 6 million years ago. Around 2 million years ago, some evolved into the genus *Homo*, characterized by an increase in brain size, a reduction in the size of the face and teeth, and an increased reliance on stone tool technology. One species, *Homo erectus*, was the first human ancestor to spread beyond Africa, migrating into parts of Southeast and East Asia, and to the easternmost fringes of Europe. Between 800,000 and 200,000 years ago, some populations of early humans evolved a brain size that was almost the same size as modern humans, but still had a relatively large face and a differently shaped skull. Many anthropologists place these finds in the species *Homo heidelbergensis*, which occupied parts of Europe, Africa, and perhaps Asia. *Homo heidelbergensis* appears to give rise to a group of early humans known as the Neandertals in parts of Europe and the Middle East.

Anatomically modern humans, with a large brain, rounded skull, reduced face, and a distinct chin, appear over the past 200,000 years. The distribution of fossils that can be classified as modern humans shows that they appeared first in Africa, with representative fossils found at 160,000 years ago (White et al. 2003) and 200,000 years ago (McDougall et al. 2005), well before the appearance of modern humans elsewhere. There appears to have been a brief expansion of modern humans into the Middle East at about 90,000 years ago, followed by a major dispersion around 60,000–70,000 years ago, with modern humans spreading into Australia, East Asia, and Europe.

These time frames indicate that modern humans appear first in Africa and, as they spread out throughout the rest of the world, they would have encountered earlier humans (such as the Neandertals) that were already there. This raises the question of possible gene flow between modern and other human populations. Anthropologists have proposed two models that focus on this initial African origin and the issue of gene flow outside of Africa. The first is the African replacement model, which proposes that modern humans completely replaced earlier humans without any interbreeding. Under this model, other humans such as the Neandertals became extinct without contributing anything to our ancestry. An alternative view, the assimilation model, proposes that modern humans interbred with other human populations outside of Africa. Here, groups such as the Neandertals would have contributed something to our ancestry even though as a distinct population they have become extinct (Smith et al. 2005; Relethford 2008a).

A. Genetic Consequences of an African Origin

For the moment, we will put aside the question of possible interbreeding and focus on the African origin of modern humans. From a genetic perspective, we are interested in seeing to what extent data on genetic variation in living humans are consistent with the African origin that we see from the fossil record, and how an African origin explains certain features of genetic variation in our species. Numerous studies of coalescent trees based on different DNA markers have shown that most (but not all) gene trees have roots in Africa, consistent with our African origin (Garrigan and Hammer 2006). In addition, Relethford (2011) lists five characteristics of genetic variation in living humans that relates in whole or in part to our African origin:

1. Low levels of genetic variation relative to other species
2. Higher levels of genetic variation in Africa
3. Genetic diversity declines with distance out of Africa
4. Low levels of genetic differentiation between geographic regions
5. A strong global correlation between genetics and geography

Comparative studies of DNA variation show that the human species has less genetic diversity than in chimpanzees, gorillas, and orangutans (Garrigan and Hammer 2006; Hawks 2009), even though we are spread out over the entire planet, whereas the great apes have smaller, more restricted populations. Our lower levels of diversity today, despite our large current population size, suggest that the human species underwent an evolutionary bottleneck in the past, leading to increased genetic drift and loss of genetic diversity. One reason for a bottleneck could have been a small founding group of modern humans that first expanded out of Africa (Garrigan and Hammer 2006).

An African origin also ties in with the observation that diversity in DNA markers is higher in Africa than in other geographic regions, such as Asia or Europe (Relethford 2008a). One possibility for higher levels of African diversity is a model where modern humans existed in Africa for a long time before a small founding population expanded out of Africa. Mutations accumulate over time, leading to higher levels of genetic diversity in Africa. When populations expanded out of Africa, some of this diversity would be lost as a result of the founder effect. This scenario also fits the observation that DNA diversity outside of Africa is often a subset of the diversity seen within Africa in terms of number and frequency of different alleles (Tishkoff and Gonder 2007). It is also possible that higher levels of genetic diversity in Africa reflects to some extent larger population size over much of the past (Relethford and Jorde 1999), resulting in less genetic drift.

The hypothesis that the expansion out of Africa led to a reduction in genetic diversity also receives support from the observation that genetic diversity declines with increasing geographic distance away from Africa (e.g., Ramachandran et al. 2005). The best way to explain this geographic pattern is by focusing on a series of founding events. Here, we imagine a small founding population leaving Africa, resulting in an initial reduction in genetic variation because of genetic drift. If this population then increases in size until another small founding group splits off, then the founder effect would lead to a further reduction in genetic variation. Over time, the continued repetition of founding, growth, and further founding would cause the geographic decline in genetic diversity from the original source (Africa) that we see today.

The fourth observation regarding our species' genetic variation is the fact that the level of genetic differentiation (F_{ST}) among geographic regions is relatively low. Numerous studies have shown that F_{ST} between geographic regions, such as sub-Saharan Africa, Europe, and East Asia, is around 0.10 (Madrigal and Barbujani 2007). This relatively low level of differentiation is consistent with a relatively recent origin of modern humans (so that equilibrium of F_{ST} has not been reached), and/or high rates of long-distance migration in recent human evolution. F_{ST} is an aggregate measure that reflects a number of possible causes, but the various possibilities are consistent with the evidence of an African origin.

Finally, numerous studies have shown a strong global correlation between genetic and geographic distances (Cavalli-Sforza et al. 1994; Relethford 2004; Ramachandran et al. 2005). Human populations are typically the most genetically similar to their neighbors. The greater the distance between populations, the greater the genetic difference. Furthermore, the correlation between genetics and geography applies to geographic distances that factor in the likely routes of population dispersion and migration. For example, when we look at likely travel routes between East Asia and the New World, we do not measure the straight-line distance across the Pacific Ocean, but we instead measure the distance along the known route from East Asia up to Siberia and then down into the Americas (this will be discussed in a case study later in this chapter). The correlation between of genetics and geography may reflect in part the series of founding events associated with the spread of modern humans across the planet. Populations geographically close to one another tend to share common ancestry and history. Of course, migration between neighboring populations (isolation by distance) could also have some impact.

The review above demonstrates that patterns of genetic variation in living humans agree with the evidence from the fossil record of an initial African origin of modern humans. What is less clear, however, is whether there was any interbreeding with preexisting human populations outside of Africa. There is some genetic evidence for interbreeding outside of Africa, as some gene trees have non-African roots (Templeton 2005; Garrigan and Hammer 2006). Some of the clearest evidence of this possibility comes from new research on ancient DNA of earlier human populations, such as the Neandertals.

B.   The Fate of the Neandertals

The question of interbreeding is most apparent when looking at the Neandertals of Europe and the Middle East. The Neandertals were a population that arose approximately 150,000 years ago and persisted in parts of western Europe until 28,000 years ago. Physically, the Neandertals were similar to modern humans in having a large brain, but had a differently shaped skull that was long and low, with large brow ridges and a large nose and midfacial dimensions. Anthropologists have long debated the question of the relationship of the

Neandertals to modern humans, wondering whether they were a subspecies of *Homo sapiens* or a separate species, and if the latter, whether they were capable of interbreeding with modern humans. It is clear that the Neandertals and modern humans lived at the same time in Europe, although the length of this coexistence is still debated. Genetically, the question is whether any gene flow took place between Neandertal and modern human populations—was there replacement or assimilation? Stated another way, we are asking whether Neandertals contributed anything to the ancestry of modern humans.

Fossil evidence has suggested at least some interbreeding, as a number of physical traits specific to the Neandertals have been found in the earliest post-Neandertal modern human populations in Europe, and decline afterward, suggesting a pattern of assimilation into a large gene pool rather than a complete and immediate replacement (e.g., Smith et al. 2005; Trinkaus 2007). Others have suggested alternatives and argue that there was little, if any, interbreeding (e.g., Tattersall 2008).

When I was in graduate school, we used to joke that the Neandertal issue could be solved if we only had DNA. At the time, the idea of extracting DNA from fossil remains was something that existed only in science fiction stories. As is often the case, however, yesterday's science fiction is today's science. Advances in genetic methods led to the successful extraction of a small piece of mitochondrial DNA from a Neandertal fossil in 1997 (Krings et al. 1997). As noted in Chapter 1, mitochondrial DNA is a small section of about DNA over 16,000 base pairs in length that is found in the mitochondria of cells and is passed on through the mother. Although the fragment of mitochondrial DNA recovered from the Neandertal fossil was small (379 base pairs in length), it was enough to show some interesting differences from living humans. This finding was not a fluke; since then, small sections of mitochondrial DNA have been recovered on over a dozen other Neandertal fossils (Hodgson and Disotell 2008). In addition, the entire mitochondrial DNA genome has been recovered in one Neandertal fossil (Green et al. 2008). These studies show that certain distinctive features of the Neandertal mitochondrial genome have never been found in living humans (Hodgson and Disotell 2008). This finding by itself suggests that Neandertals did not contribute to the ancestry of living humans—there was no interbreeding.

However, the situation was not entirely clear when we considered only the mitochondrial DNA. Because of its haploid inheritance with no recombination, mitochondrial DNA essentially functions as a single locus. As such, there is much random variation associated with sampling error that makes resolution difficult. A consensus has been that the mitochondrial DNA evidence suggested either zero interbreeding or a small amount of interbreeding. The case for Neandertal interbreeding became stronger with another example of science fiction becoming science fact—the sequencing of a large proportion of the complete nuclear genome of Neandertals (Green et al. 2010). Comparison of this sequence with the genomes of a chimpanzee and five humans from around the world has produced some interesting results. Although very similar, genomic analysis suggests that the Neandertal and modern human lineages diverged from a common ancestor about 270,000–440,000 years ago. Under a replacement model where there was no interbreeding after this split, we should see equal differences between Neandertal

DNA and living human DNA across the world (because the dispersion out of Africa occurred after the initial split). Instead, the genomic comparisons show that Neandertal DNA is actually more similar to living Europeans and Asians than to living Africans. This suggests that some interbreeding occurred *after* the dispersion of modern humans out of Africa but before subsequent dispersions throughout Eurasia, perhaps due to overlap of modern humans and Neandertal in the Middle East. Additional genome comparisons suggest that the gene flow took place from the Neandertals into modern humans, such that between 1% and 4% of the ancestry of living Eurasians came from the Neandertals. The DNA evidence shows clearly that there was *not* complete replacement. The Neandertals *as a population* have disappeared, but some of their DNA lives on in nontrivial amounts in our species today.

The survival of Neandertal DNA may not be an unusual feature of recent human evolution. Genomic analysis has also been performed on a human finger bone dating to about 30,000–50,000 years ago from Denisova Cave in southern Siberia (Reich et al. 2010). These analyses show yet another unique genome related to both the Neandertals and modern humans. This population, referred to as the *Denisovans*, appears to be an archaic group that diverged from the line leading to the Neandertals 640,000 years ago. Comparison of the Denisovan genome with the Neandertal genome and genomes of living humans suggests gene flow from the Denisovan population into living Melanesians, but not into other human populations. Further, Reich et al. (2010) estimated that 4–6% of the ancestry of Melanesians came from the Denisovans. To add to the mystery, we do not have much information on exactly what the Denisovans looked like, as the fossil evidence from the cave is only a finger bone and a tooth.

It may seem very strange at first to think of a scenario where an archaic human population lived in Siberia but contributed genetically to Melanesians, who are far apart! It is possible that the Denisovans were part of a geographically widespread population across parts of Asia that is now extinct, but that some of these populations contributed genes to the East Asian populations that later spread into Melanesia. We are beginning to see glimpses of a complex picture of recent human evolution where dispersion out of Africa was accompanied by limited interbreeding in different times and places with populations that are now extinct (Bustamante and Henn 2010). In some parts of the world, there may have been less gene flow from archaic populations, and in other parts, there might have been more. In Melanesia, it appears that there has been gene flow from *both* the Neandertal line and the Denisovans, such that over 7% of the ancestry of modern Melanesians ultimately derives from archaic gene pools.

III. CASE STUDIES OF POPULATION ORIGINS

The remaining case studies on population origins focus on regional and local studies of population origins and histories in several different parts of the world.

A. The Peopling of the New World

In 1492, the first of Columbus's voyages led to the "discovery" of the New World by Europeans. Columbus' plan was to circumnavigate the world in order to find a quicker route to parts of Asia then referred to as the "Indies." [*Note*: The idea that Columbus was trying to prove the earth was round turns out to be a myth—people in his time and earlier had already deduced that the earth was round (Gould 1999).] Columbus made the mistake that he had arrived in the Indies, and because of this, the local people in the New World became known as "Indians." Ever since, scholars and laypeople have wondered about the initial origin of the native peoples of the Americas. Some of these suggestions included ideas that Native Americans were a lost tribe of Israel, voyagers from ancient Egypt, or survivors from Atlantis (Crawford 1998). A number of people early on recognized that northeastern Asia made the most sense as a point of origin, as the Asian and North American continents are close together in the arctic, separated by only the Bering Strait.

The traditional explanation of migration, long favored by archaeologists, is that humans in Siberia were able to move across the Bering Strait during glacial times; during an "ice age," the sea level drops, exposing the land that connects the two continents. Human hunters following animal herds would have then crossed over this "land bridge" (a bit of a misnomer, as this "bridge" was actually about 1000 miles wide, and this migration was more of an expansion into new territory over time, rather than a march across a bridge). After entering the Americas, these nomads would have been stopped by massive glaciers, except for times during which two glaciers had receded enough for humans to move southward into the rest of the Americas by passing through an ice-free corridor. This traditional model was based on the hypothesis that humans entered the New World during a time when the land bridge *and* the ice-free corridor were open, roughly 12,000 years ago. Evidence of *earlier* occupation has led to this view being questioned, and some archaeologists have suggested that some early migrants used boats to move into the New World (Nemecek 2000), and some may have travelled southward along the western coast of North America (Dalton 2003). The increasing evidence for early occupation has increased, with a number of sites suggesting dates as far back as 15,000 years (Goebel et al. 2008; Waters et al. 2011).

Regardless of the timing or routes of migration, the fact that the first Americans came from northeastern Asia is incontrovertible. Close genetic similarity is seen in a variety of physical and genetic measures. Analysis of genetic distances based on red and white blood cell markers shows that Native American populations are more genetically similar to East Asian populations than to populations elsewhere in the world, such as Europe, Africa, or Australia. Furthermore, the closest levels of genetic distance to Native American populations are with northeastern arctic populations, such as Siberians (Cavalli-Sforza et al. 1994; Crawford 1998).

The genetic connection between northeastern Asians and Native Americans is even clearer when we examine combinations of DNA markers, specifically mitochondrial DNA haplogroups. Recall from Chapter 4 that haplogroups are sets of related haplotypes that share some common mutations. Early studies of mitochondrial DNA in our species revealed over two dozen different haplogroups, many of which have relatively restricted geographic distribution (Jobling et al. 2004). Among Native Americans, there are four major haplogroups, labeled A, B, C, and D. These haplogroups are *not* found in Africa, Europe, or South Asia—instead, they are found only in East Asia. The simplest explanation for this distribution is that the first Americans came originally from East Asia.

Among North American Indians, a fifth haplogroup, X, has been found, and this haplogroup is also found in parts of Europe. Could the presence of haplogroup X indicate recent European admixture, or perhaps some earlier contact with Europeans before Columbus' time? Although the initial lack of haplogroup X in East Asian samples was suggestive of European contact, later studies confirmed that this haplogroup *does* exist in Siberian populations (Derenko et al. 2001).

The few examples above are given simply to illustrate the type of genetic information that has been used to confirm the East Asian origin of Native Americans, and are not exhaustive of the full extent of genetic data. Continuing questions about Native American origins deal with the question of the number of migrations into the New World (many genetic studies favor a single migration event, but this has not been accepted universally) and the timing of entry into the New World.

B. Genetics and the Spread of Agriculture

Humans began domesticating plants and animals about 12,000 years ago in the Middle East. Over time, agriculture independently developed in other parts of the world, including Africa, East Asia, and the Americas. An exception to the general rule of an independent origin of agriculture was Europe, where archaeological evidence shows that agriculture there came out of the Middle East. The spread of agriculture into Europe began about 9000 years ago in Iraq and Turkey, and moved in a northwest direction over the next 3000 years (which works out to about 1 km per year). The process of a new behavior spreading from one population to another is known as *diffusion*. There have been two schools of thought about exactly *how* the diffusion of agriculture into Europe took place. One model, known as *cultural diffusion*, states that agricultural methods spread from group to group as part of the cultural contact between neighbors. An alternative model, known as *demic diffusion*, proposes that groups of farmers physically moved out of the Middle East into Europe, bringing this new behavior with them (Ammerman and Cavalli-Sforza 1984).

The difference between these models is whether the farmers actually moved. Under cultural diffusion, no people actually moved, and the spread of agriculture represented the spread of an idea, passed from group to group. Under demic diffusion, the farmers are expanding into Europe, bringing this new technology with them, and interbreeding with the existing populations in Europe. Elsewhere (Relethford 2003), I use a simple analogy to contrast these models. Consider football players spread out every 10 yd (yards) across a football field, with the player on one end of the field having the football. There are two ways that the football can cross the field. Analogous to cultural diffusion, each player can pass the football to the next person 10 yd away. This process continues from player to player until the ball has moved across the entire field but the players have remained in place. Another way of moving the ball, analogous to demic diffusion, is for the person with the ball to walk across the entire field while carrying the ball. According to both models, the football moves across the field, but under the first model the players remain where they are, and in the second model, they move. In terms of the diffusion of agriculture, the cultural diffusion model argues that farming methods diffused, but not the farmers. Under the demic diffusion model, the farmers move, taking farming methods with them.

It would seem that either model produces the same result—agriculture spreading into and across Europe. How can we distinguish between these models and figure out whether the farmers moved? One way that this problem has been examined is with genetic data. A key difference between the two diffusion models is that the demic diffusion model predicts that the actual populations move across space, and would interbreed with the populations that they encountered. If this happened, then there would be a gradient in allele frequencies, known as a **cline**, in the same direction as the spread of agriculture. Here, the allele frequency would increase (or decrease—the direction is irrelevant) from the Middle East in a northwest direction into Europe. On the other hand, if the spread of agriculture was through cultural diffusion, then the farmers did not move into Europe and there would be no cline.

Cavalli-Sforza and colleagues tested these hypotheses by looking at the spatial distribution of alleles from a large number of genetic markers (Menozzi et al. 1978; Cavalli-Sforza et al. 1993, 1994). The approach used was to map allele frequencies to see if there was a southeast to northwest direction coinciding with the spread of farmers out of the Middle East into Europe. Such maps are very similar to the temperature maps that you see in newspapers, where the clines correspond to different temperatures. Rather than looking at dozens of individual allele frequency maps, Cavalli-Sforza and colleagues developed a way of looking at common directions in sets of allele frequencies using a statistical method known as *principal components analysis*. This method produces a small number of components that represent common and unique patterns of variation in dozens of allele frequencies. They found that the first principal component (which accounts for the majority of spatial variation in allele frequencies) showed a definite correspondence with the spread of agriculture in Europe. This cline shows that a large number of alleles have distributions shaped by farmers moving across Europe, which supports the demic diffusion model.

An interesting sidenote is that the other principal components showed evidence of *other* past population movements, such as an expansion from eastern to western Eurasia, corresponding to the spread of peoples that had domesticated horses in Russia and then moved east. Other components showed evidence of other prehistoric movements. We have to remember that the current state of genetic variation has been shaped by multiple events. The spread of farmers out of the Middle East was an important migration event, but it was not the only one.

Although classical genetic marker studies have supported the demic diffusion model, implying a major genetic input from the Middle East, results from DNA marker analysis have not been as clear. Reviewing studies of both mitochondrial DNA and Y-chromosome DNA markers, Richards (2003) suggested that the total genetic contribution from incoming farmers was likely less than 25%, and that the genetic consequences of the spread of agriculture was minor. Pinhasi and von Cramon-Taubadel (2009) noted that different sets of data have produced different results. In general, demic diffusion has been supported by analyses of nuclear

DNA markers and cranial measurements, while mitochondrial DNA markers suggest less of an effect, and Y-chromosome DNA markers give mixed results. However, ancient DNA analysis from skeletons of a prehistoric farming society in Germany roughly 7000 years ago shows clear genetic affinity with Middle Eastern populations, supporting the demic diffusion model and a significant genetic component of ancestry from the Middle East (Haak et al. 2010). Although the population dynamics shown in all of these studies are more complicated than a simple model, the overall results support demic diffusion.

C.   The Colonization of Polynesia

Humans have a long history of traveling over short distances of water. As noted earlier in this chapter, some of the first Americans may have traveled via boats into the New World. We also know that the first humans to reach Australia some 40,000–60,000 years ago must have been able to cross some water to get there. The most extensive early sea voyages, however, took place with the colonization of islands in the Pacific Ocean by early Polynesians. The name "Polynesia" literally translates as "many islands," and refers broadly to a large triangular area of the Pacific Island whose vertices are New Zealand in the south, Hawaii to the north, and Easter Island to the east. Polynesia is one of three geographic and culture areas of the Pacific Ocean; the other two are Melanesia (meaning "dark islands" after the dark skin color of the native populations) and Micronesia (meaning "small islands").

Archaeological and genetic evidence shows that the first Polynesians arose in southern or southeastern Asia and then expanded from west to east across the Pacific. This expansion has been traced through the archaeological remains of a population known as the *Lapita culture*, named after a particular type of pottery. The Lapita were farmers who also relied heavily on fishing for their subsistence. The dates for Lapita pottery show a clear west–east expansion out of South/Southeast Asia; remains have been found showing colonization near New Guinea about 3600 years ago, at Samoa about 2200 years ago, at New Zealand about 1000 years ago, and at Hawaii and Easter Islands about 1500 years ago. The first Polynesians crossed over the ocean in outrigger canoes, designed to maintain balance on the ocean's waves (Diamond 1997).

Analysis of classical red blood cell and white blood cell markers shows that Polynesians today are most similar to South and Southeastern Asian populations, and more distant from populations elsewhere in the world (Cavalli-Sforza et al. 1994; Relethford 2003). The affinity with South/Southeastern Asia is consistent with the archaeological evidence. An alternative view had been proposed many years ago by the Norwegian explorer Thor Heyerdahl, who suggested that the inhabitants of Easter Islands actually came from South America rather than Asia. The genetic data refute this theory, as Polynesians are more similar to South/Southeastern Asian populations than to Native American populations (Relethford 2003).

Although the ancestors of Polynesians were the first people to move into the far areas of the Pacific Ocean, they did have to pass near occupied parts of Melanesia, where humans had been living for 35,000 years, on their way eastward from Asia. The obvious question is whether there was any interbreeding between the first Polynesians and Melanesians. Do living Polynesians have any Melanesian ancestry? Two primary models have been used to explain the west–east expansion into Polynesia, and each has its own genetic implications. The "express train" model, suggested by Diamond (1988, 1997) is aptly named, as he argues that expansion out of Asia took place very quickly, moving eastward past Melanesia like an express train with no stops. Genetically, this would mean that little, if any, interbreeding took place, and there should be little or no evidence of Melanesian ancestry in living Polynesians. An alternative model, dubbed the "slow boat" model by Kayser et al. (2000), suggests slower movement through Melanesia and consequently more gene flow into the Polynesian population.

Studies of genetic distance based on red and white blood cell markers suggest little Melanesian ancestry (e.g., Cavalli-Sforza et al. 1994), but resolution of ancestry is difficult for this region. Haploid DNA markers give different results for mitochondrial DNA and Y-chromosome DNA. The frequency of mitochondrial DNA haplotypes shows a clear geographic pattern that correlates with an expansion out of South/Southeast Asia into the Pacific as predicted by the express train model. Here, there is no evidence of Melanesian admixture (e.g., Redd et al. 1995; Sykes et al. 1995). However, analysis of Y-chromosome DNA markers has shown a different picture. Here, some haplotypes have been found in both Polynesian and Melanesian populations suggesting appreciable Melanesian admixture and support for the slow boat model (Kayser et al. 2000). Other studies of Y-chromosome DNA have also provided support for Melanesian ancestry (e.g., Capelli et al. 2001; Underhill et al. 2001).

Conflicting results from mitochondrial DNA and Y-chromosome DNA could reflect sex differences in ancestry (different ancestry through the maternal and paternal lines). Of course, one problem with these haploid markers is that they are inherited as a single unit and act as a single locus, which introduces a lot of sampling error. In addition, these markers have lower effective population sizes (because they are haploid), which increases genetic drift, making determination of ancestry more difficult. Studies of a larger number of nuclear DNA markers have provided resolution for the debate over Polynesian ancestry. Friedlaender et al. (2008) examined a very large number of DNA markers (687 STR loci and 203 insertion/deletion loci) from 41 Pacific populations and found that Polynesian populations clustered with East Asians and South Asians but not with Melanesians. They conclude that the first Polynesians moved relatively quickly through Melanesia with only a small amount of admixture. In another study, Wollstein et al. (2010) examined roughly 1 million (!) SNP loci and came up with similar conclusions. Their analysis suggested that roughly 87% of the ancestry of the first Polynesians came from East Asia. In both cases, these studies support the express train origin model.

D.   The Origin of the Irish Travelers

There are a small number of nomadic groups living in Europe, including the Roma (often referred to as "gypsies") of central and eastern Europe. One of these itinerant groups are the Irish Travelers (formerly known as "Tinkers"), who make up a very small percentage (<1) of the population of Ireland. The Travelers are a social group that moves around the countryside performing odd jobs, seasonal labor, and scavenging scrap metal, among other activities (Gmelch 1977). There have been several ideas about the ancestral origin of the Irish Travelers. Some of these hypotheses focus on the idea that the Travelers are the descendants of Irish who were displaced from their land, becoming an isolated social group over time. Another idea is that they represent a mixture of Irish and Romany gypsies. The difference between these two sets of ideas can be tested using genetic data.

Crawford and Gmelch (1974) computed genetic distances between a sample of Irish Travelers and a number of European and Asian populations based on red blood cell genetic markers. They found that the Travelers were most similar to other Irish populations, and more different from other populations, including several gypsy samples. They concluded that the genetic data supports the hypothesis of an Irish origin for the Irish Travelers. Other analyses of the same data using different comparative samples find the same result (Croke et al. 2000; North et al. 2000). The cultural similarity to some aspects of gypsy lifestyle is coincidental and does not reflect ancestry. As is often the case with human populations, culture is independent of genetics.

It is likely that genetic drift has had an important effect on genetic variation in the Traveler population. The small size of the Traveler population and the high variance in fertility (Crawford and Gmelch 1974) suggest a relatively low effective population size, which would increase the likelihood of drift. Drift is also suggested from studies of a metabolic disorder known as *transferase-deficient galactosemia* that show much higher frequencies among the Irish Travelers than among other Irish (Murphy et al. 1999). DNA analysis showed that all of the cases among the Travelers were due to a specific mutation known as Q188R, which also accounted for 89% of the mutant alleles for the cases among the non-Travelers. Screening of the overall population shows that the allele frequency of the Q188R mutation is much higher among the Travelers (0.046) than among the non-Travelers (0.005). It appears likely that this mutation arose in Ireland and attained an elevated frequency among the Travelers because of genetic drift. Some of this difference might be due to initial founder effect, and some due to continued genetic drift in subsequent generations.

E. Admixture in African-Americans

During Colonial times, hundreds of thousands of Africans were enslaved and brought forcibly to the United States as part of the slave trade. Most came from western Africa and western central Africa. Starting in 1619, the slave trade grew and the importation of enslaved Africans was widespread, peaking at the end of the eighteenth century and the beginning of the nineteenth century (Parra et al. 1998). Over the centuries, there has been gene flow from those of European ancestry into the gene pool of African Americans. Some of this gene flow occurred before the prohibition of slavery, generally when European men would mate with enslaved African-American women. More recently, there has been gene flow from matings between European-Americans and African-Americans following the relaxation of social barriers to interracial marriage.

European gene flow into African American populations is an example of the process of admixture as described in Chapter 8, and genetic marker data can be used to estimate the accumulated European ancestry in African-Americans. Some studies have used the simple model of two parental populations (Europe, Africa) described in Chapter 8, and some have looked at more complex models that take into account possible admixture from Native American sources. The earliest studies soon found that the amount of European admixture varied geographically (Chakraborty 1986), such that there is no single or simple answer to the question of how much European ancestry exists in African Americans. Genetic history is not the same in all populations.

The variation in European ancestry is shown clearly by a comprehensive study of African-American genetics conducted by Esteban Parra and colleagues (Parra et al. 1998). Here, they examined genetic markers from nine loci that exhibited large differences between Europeans and Africans, and estimated admixture proportions in 10 different African-American populations in the United States. They found that the amount of accumulated European ancestry ranged from 12% in Charleston, South Carolina to 23% in New Orleans, Louisiana.

These estimates were based on autosomal genetic markers. Parra et al. (1998) also examined admixture estimates based on mitochondrial and Y-chromosome DNA markers, which allow us to look at the different evolutionary histories of the female and male lines. The mitochondrial DNA analysis showed European ancestry ranging from 0% in Detroit, Michigan to 15% in Baltimore, Maryland. The estimates from Y-chromosome DNA markers ranged from 9% in Houston, Texas to 47% in New Orleans. More revealing is the fact that in each of the nine populations studied (Y-chromosome DNA was not available in Charleston) the amount of European ancestry was greater in the male line than in the female line. These results suggest that over the past few centuries there have been more European genes coming from the mating of European-American males with African-American females than from the mating of European-American females with African-American males. As noted above, in preslavery times the majority of interracial matings were probably between a male of European descent and an enslaved female of African descent. This pattern appears to be reflected in the mitochondrial and Y-chromosome DNA analyses. In the last 40 years, there have been more marriages between a black husband and a white wife than between a white husband and a black wife, with a ratio of about 2:1 in the 1990s (US Census web page, http://www.census.gov/population/socdemo/ race/interractab1.txt). This recent demographic shift will result in an increase in the maternal component of European ancestry in African-Americans, but it has not yet had a noticeable effect on the genetic history of past events.

Keep in mind that all of these estimates are based on the genetic makeup of an entire population, and the admixture proportions of a population do not necessarily apply to every person in that population. Each person has his or her own, often unique, genetic history, and our population estimates are simply statistical aggregates of these histories. An estimate of 15% European ancestry for a sample does *not* mean that every person in that population has 15% European ancestry. As with population estimates, individual estimates of ancestry can vary considerably.

A good example of individual variation in ancestry is Parra et al.'s (2001) study of European ancestry in several African-American populations in South Carolina. Here, they estimated admixture for individuals as well as for the entire population. The estimate for Columbia as a whole was 18% European ancestry, but there was considerable variation in individual ancestry; the majority of the individuals had less than 10% European ancestry, while roughly 7% had more than 50% European ancestry. However, among the more culturally isolated Gullah people of South Carolina, over 80% of them have less than 10% European ancestry, and no one has more than 50% European ancestry.

Ancestry varies from population to population, and from individual to individual. Any attempt to characterize the genetic history of *all* African-Americans (or any human population, for that matter) by a single number is futile. The genetic studies of Parra and colleagues also shows how we need to keep a distinction between one's genetic ancestry and their cultural identity. Some of the individuals in the Columbia study actually have more European ancestry than African ancestry, yet all are self-defined as African-American or "black." The difference is that the admixture estimate deals with genetic ancestry and the self-definition refers to cultural identity. Another example to consider is the case of current (2011) US President Barack Obama. His mother was of European descent and his father of African descent. From the perspective of genetic ancestry, he can be described as "half-black" and "half-white." Yet, from a cultural perspective of how he describes himself and how others describe him, he is labeled African-American or black. As noted by Marks (1994), such individuals can simultaneously be described as "half-black" and "black." How can you be something and half of something at the same time? This apparent conflict is easily resolved by noting that "half-black" is a statement of genetic ancestry and "black" is a statement of cultural identity. As always, when we study *human* population genetics, we need to take an anthropological perspective that encompasses both biology and culture.

IV. SUMMARY

Studies of neutral genetic variation consider the interaction between and effects of mutation, genetic drift, and gene flow on patterns of variation both within and between populations. Anthropologists and human geneticists have long studied both the genetic *structure* of populations and the genetic *history* of populations using neutral genetic markers. Studies of human population structure focus on patterns of mate choice and how geographic, environmental, and cultural factors influence the genetic makeup of individuals within populations and the pattern of genetic similarity between populations. This chapter has considered several case studies of human population structure in New Guinea, northern Europe, and Argentina. These studies illustrate some (but not all) of the ways in which genetic variation is structured in human populations.

Studies of the genetic history form an important part of human population genetics research, ranging from the history of our entire species to the history of large geographic/cultural units to the history of local populations. A continuing focus of population history studies is the origin and dispersion of modern humans.

The African origin of modern humans (*Homo sapiens*) some 200,000 years ago is clear from the fossil record, as well as the dispersal of modern human populations across the world. This African origin has left a noticeable imprint on our species' genetic diversity. Additional case studies in this chapter also outline examples of regional and local genetic history.

GLOSSARY

absolute fitness   A measurement of fitness comparing the actual number of individuals by genotype at two different times, such as generation to generation, or birth and adulthood.

admixture   Gene flow between distant and previously separated populations.

allele   An alternative form of a gene or DNA sequence. For example, the gene that determines human blood type ABO has three alleles (different forms): *A*, *B*, and *O*.

allele frequency   The proportion of a given allele among all alleles. For example, if a population has 35 *A* alleles and 15 *a* alleles in the gene pool, the frequency of the *A* allele is $\frac{35}{50} = 0.7$.

among-group variation   Genetic differences between individuals *within* the same population.

ancestry-informative markers (AIMs)   Genetic markers used in admixture analysis whose allele frequencies in parental populations are very different, thus increasing the accuracy of admixture estimation.

assortative mating   Mating choice based on phenotypic similarity. Typically, humans practice positive assortative mating, which is a preference for mating with someone who is phenotypically similar (negative assortative mating would be a preference for mating with someone who is phenotypically different).

autosomes   A chromosome in the nucleus that is not a sex chromosome. Humans have 23 pairs of chromosomes. One of these contains the sex chromosomes, and the other 22 pairs are referred to as *autosome pairs*.

balancing selection   Selection for the heterozygote, which leads to an equilibrium allele frequency >0.0 and <1.0. Instead of one allele reaching a frequency of 1.0 or 0.0, a balance is achieved.

base pair (bp)   A nucleotide on a strand of DNA or RNA that is paired with its complement on the other strand (the chemical base A pairs with T, and the base

Human Population Genetics, First Edition. John H. Relethford.
© 2012 Wiley-Blackwell. Published 2012 by John Wiley & Sons, Inc.

C pairs with the base G). The length of DNA/RNA sequences is measured in base pairs. For example, the total length of the human genome, including all 23 chromosome pairs, is over 3 billion base pairs in length.

between-group variation   Genetic differences between populations.

bottleneck   A dramatic reduction in population size that results in an increase in genetic drift.

bp   An abbreviation of *base pair*.

breeding population size   The number of individuals in a population that are capable of reproducing, typically defined as those of reproductive age.

census population size   The total number of individuals in a population, which includes people of reproductive age as well as those that are too young or too old to reproduce.

chi-square test   A statistical test that can be used to compare observed and expected frequencies. In Appendix 2.2, a chi-square test is used to determine whether there is significant deviations of observed genotype numbers from those expected under Hardy–Weinberg equilibrium.

chromosome   A long strand of DNA. Chromosomes come in pairs; humans have 23 pairs of chromosomes.

cline   A gradient in allele frequencies, typically associated with gene flow.

coalescent theory   Population genetic models that focus on the probability that members of a population share a common ancestor in the past. Unlike classical models of population genetics that examine genetic changes forward in time, coalescent models track ancestry backward into the past. Coalescent theory shows how any two copies of an allele share a common ancestor in the past; that is, they coalesce to a common ancestor.

codominant   Term describing arrangement in which alleles in a heterozygote both affect the phenotype. Neither allele is dominant or recessive.

consanguineous marriage   Marriages where bride and groom are closely related, such as a marriage between cousins.

culture   At the most basic level, this is shared, learned behavior. Studies of human population genetics find that culture often plays an important role in human microevolution, and must always be considered when looking at human genetic diversity.

deletion   A mutation resulting in the deletion of base pairs.

diploid   Term referring to the presence of two copies of an allele, inherited from two parents, at a given locus. The vast majority of human genotypes represent diploid inheritance.

directional selection   Selection of a quantitative trait where one extreme phenotypic value is selected for and the other extreme phenotypic value is selected against. This type of selection moves the average phenotypic value over time in one direction or the other (larger or smaller, depending on the direction of the selection).

dominant   Term referring to an allele that always shows its effect, even if only one copy is inherited. Dominant alleles are dominant over recessive alleles.

effective population size   The breeding population size in an idealized population where a number of conditions, such as equal sex ratio and constancy in population size, apply. Formulas for effective population size allow deviations from these assumptions to be taken into account in estimating the true breeding size of the population.

endogamy   When a mate is chosen from within a group, as contrasted with exogamy.

electrophoresis   A method of laboratory analysis that uses electricity to separate proteins or strands of DNA in order to indentify different genotypes.

equal and additive effects model   A model of the polygenic inheritance of a quantitative trait, where each locus contributes equally to the phenotype.

evolutionary forces   The four mechanisms that can cause a change in allele frequency from one generation to the next: mutation, natural selection, genetic drift, and gene flow.

exogamy   When a mate is chosen from outside a group, as contrasted with endogamy.

extinction   In the context of genetic drift, when an allele is lost (reaches a frequency of $p = 0.0$).

fitness   The probability of and individual with a given genotype surviving and reproducing.

fixation   In the context of genetic drift, when an allele reaches a frequency of $p = 1.0$.

fixation index   A measure of the proportional reduction in heterozygosity relative to the amount expected under Hardy–Weinberg equilibrium. Fixation indices can be used to study effects of inbreeding and genetic drift.

founder effect   A reduction in genetic diversity due to genetic drift when a new population is made up of a small number of founders.

gamete   A sex cell.

gene   In strict usage, a DNA sequence that codes for a functional product, such as a protein. In broader usage, *gene* is sometimes used to refer to any identifiable section of DNA, regardless of whether it has functional significance (= genetic marker).

gene flow   A change in allele frequency caused by the movement of genes from one population into another. Gene flow is not always synonymous with migration.

genetic drift   A random change in allele frequency from one generation to the next due to sampling; by chance, the next generation may not have the exact same allele frequency. The expected effect of drift is greater when population size is small.

genetic marker   A DNA sequence that has a known location on a specific chromosome and can be identified.

genome   The total amount of DNA of an individual.

genotype   The genetic makeup of an individual at a given locus defined by what has been inherited from both parents.

genotype frequency   The proportion of individuals that have a given genotype. For example, if there are 18 individuals with genotype AA out of a total of 50 individuals, the frequency of the AA genotype is $\frac{18}{50} = 0.36$.

haplogroup   A set or related haplotypes that share common mutations.

haploid   Where there is one copy of an allele present at a given locus. One example is in sex cells, which contain only one chromosome from each pair. Mitochondrial DNA is an example of haploid inheritance; you received only mitochondrial DNA from your mother.

haplotype   A combination of alleles that are inherited as a single unit.

Hardy–Weinberg equilibrium   The mathematical relationship between allele frequencies and the expected genotype frequencies in the next generation given the assumptions of random mating, no mutation, no selection, no drift, and no gene flow. Under Hardy–Weinberg equilibrium, allele frequencies and genotype frequencies remain constant over time, thus providing a baseline condition for predicting evolutionary change.

heterozygosity   The proportion of heterozygotes in a population. When there are only two alleles, the heterozygosity is $H = 2pq$.

heterozygous   Term used to describe a genotype where the two alleles are different.

hierarchical population structure   A model of genetic variation where the total population (T) is made up of a number of subpopulations (S) that are connected by gene flow. The measure F_{ST} is used to estimate the proportion of genetic variation between subpopulations relative to the amount of genetic variation in the total population.

homozygous   Term used to describe a genotype where both alleles are the same.

identity by descent   When two alleles are identical because they were both inherited from a common ancestor. Identity by descent contrasts with identity by state.

identity by state   When two alleles are identical but did not come from a common ancestor. Identity by state contrasts with identity by descent.

inbreeding   Mating between close biological relatives, usually cousin marriage or closer.

inbreeding coefficient   The probability that an inbred individual has two alleles at a given locus that are identical because of inheritance from a common ancestor (i.e., identity by descent).

incest   Mating between very close relatives that is most often prohibited by society, such as parent–child and brother–sister mating.

indel   The name for a class of mutations that result in insertion or deletion of base pairs.

infinite alleles model   A model where each new mutation is unique, creating a new allele, and not the same as any previous mutation.

infinite sites model   A model where a mutation at any given nucleotide site is considered unique, and happens only once. Because the genome is so large, the fact that the probability of a mutation at any given site is very low means that we can treat the total genome as infinite in length.

insertion   A mutation resulting in the insertion of base pairs.

inversion   A mutation resulting in an entire section of a chromosome ending in reverse order.

island model   A simple model of gene flow that describes the effect of one-way gene flow from a "continent" into an island, and how the allele frequency of the island becomes more similar to that of the continent over time.

isolation by distance   A model of the interaction between gene flow and genetic drift that predicts that genetic similarity between pairs of populations decreases as the geographic distance between pairs of populations increases.

isonymy   Marriages between a man and woman who have the same last name. The frequency of isonymous marriage is typically 4 times that of the mean inbreeding coefficient in a population.

kin-structured migration   When migrants are related rather than being a random sample of the source population. Kin-structured migration can reduce the typical homogenizing effect of gene flow to some extent.

linkage   When alleles on the same chromosome are inherited together because they are in close proximity to each other.

linkage disequilibrium   When two alleles are found together in a gamete more or less frequently than expected under Hardy–Weinberg equilibrium. (Also known as *gametic disequilibrium*.)

loci   The plural form of *locus*.

locus   A specific location of a gene or other DNA sequence on a chromosome. For example, the gene responsible for human ABO blood type is referred to as the ABO locus (and is found on chromosome number 9).

macroevolution   Long-term evolutionary change, typically over many millennia, and focusing on large-scale changes, such as the origin of species.

major gene model   A model of polygenic inheritance where one locus has a major effect on a phenotype relative to the other loci.

mean fitness   The average fitness of all genotypes, computed as the weighted average fitness, where weighting is by genotype frequency after selection.

meiosis   The formation of sex cells, where each sex cell contains one of each chromosome pair. In bisexual organisms such as humans, each parent contributes one of each chromosome pair. Thus, you pass on only half of your genome.

Mendel's law of independent assortment   Also known as *Mendel's second law*, this law states that chromosome pairs segregate independently of each other.

Mendel's law of segregation   Also known as *Mendel's first law*, this law states that chromosome pairs segregate during meiosis such that a sex cell contains only one allele from any chromosome pair.

microevolution   Evolutionary change over a relatively short number of generations, typically studied by focusing on allele frequency change from one generation to the next.

migration matrix analysis   A mathematical method that uses information on migration rates between populations and on population size to predict patterns of genetic similarity within and between populations under a model of equilibrium between gene flow and genetic drift.

mismatch   The case in which corresponding nucleotides on two different DNA sequences are not the same.

mitochondrial DNA (mtDNA)   A small amount of DNA (16,569 bp in length in humans) located in the mitochondria of the cell. Mitochondrial DNA is inherited only through the mother, and is an example of haploid inheritance.

mitosis   The replication of chromosomes in body cells, where each cell produces two identical cells.

most recent common ancestor (MRCA)   In coalescent theory, all alleles coalesce in the past, ultimately coalescing to a single allele. The ancestor that has this allele is known as the most recent common ancestor.

mutation   A random change in the genetic code. Mutations introduce new alleles into a population and are the ultimate source of all genetic variation. Mutations must occur in sex cells to have an evolutionary impact.

natural selection   A change in allele frequency that occurs when there is differential survival and/or reproduction among genotypes.

nearly neutral theory of evolution   An extension of the neutral model that allows for some weak selection rather than complete neutrality. Under weak selection in a small population, genetic drift will have the major impact on the distribution of allele frequencies. In large populations, selection has a greater impact.

neutral theory of evolution   A model that predicts evolutionary change based on genetic drift operating on neutral mutations (those that do not affect fitness).

nucleotide diversity   A measure of genetic variation computed from DNA sequence data, which is the average proportion of differences between all DNA sequences in a sample.

phenotype   The physical manifestation of a genotype. For simple traits, the phenotype depends on whether the genotype is homozygous or heterozygous, and whether the latter whether the alleles are codominant or dominant/recessive. For more complex traits, the phenotype reflects what has been inherited (genotype) and the influences of the *environment*, broadly defined as everything other than genetic inheritance.

phylogeography   Methods of analysis that look at the geographic distribution of genetic lineages, useful in determining the origin and geographic dispersion of a species.

point mutation   The simplest form of a mutation, which involves a change in a single base, such as from C to T.

polygenic   Term describing inheritance where the genotype reflects two or more loci.

polymorphism   Literally translated as "many forms," a genetic polymorphism is a locus where there is at least two alleles, and both have frequencies greater than 0.01.

population   A group of individuals from which most mates are chosen.

population genomics   The term sometimes used to refer to new advances in population genetics that rely on comparison of entire genomes.

population history   The study of how historical events have affected patterns of genetic variation within and between populations.

population structure   The study of how patterns of mate choice and population composition affect patterns of genetic variation within populations and between subdivisions of a population. Some areas of interest in human population structure include the effects of geographic, religious, and linguistic subdivision on genetic variation.

potential-mates analysis   Methods that study cultural and demographic influences on mate choice by using computer simulation to define likely mates and then compare characteristics with observed data to determine random and nonrandom patterns of mating.

Punnett square   A method that uses a simple table to predict the possible genotypes of offspring resulting from parents whose genotypes are known.

quantitative genetics   The subfield of genetics that looks at quantitative traits, such as height, cranial shape, and skin color.

quantitative traits   Complex physical traits whose phenotype varies continuously, rather than falling into a small number of discrete categories. Quantitative traits are affected by both genetic and environmental factors.

recessive   An allele whose effects are masked by a dominant allele, such that two copies of a recessive allele are needed to show its effects.

recombination   During meiosis, sections of DNA can cross over from one chromosome to the other chromosome in a pair of chromosomes. As a result, a given chromosome could contain parts of both maternal and paternal chromosomes.

relative fitness   A measurement of fitness one the genotype is assigned a value of 1.0 and the fitness values of other genotypes are defined relative to that genotype. Often, but not always, the genotype with the highest absolute fitness is assigned the relative fitness of 1.0.

restriction fragment length polymorphisms (RFLPs)   Identification of different alleles based on the length of DNA strands that are cut into fragments by restriction enzymes that are targeted to specific DNA sequences.

selection coefficient   The probability that an individual does *not* survive and reproduce. The selection coefficient for a genotype (s) is the opposite of the relative fitness of that genotype (w): $s = 1 - w$.

selective sweep   When natural selection reduces genetic variation at a locus, there is also a reduction in molecular variation in neighboring DNA sequences, even when they are neutral, because they are "swept along." This process is also known as "hitchhiking."

sexual selection   The hypothesis that some traits have evolved because of competition between members of the same sex for mates.

short tandem repeats (STRs)   Short repeated sequences of DNA typically ranging in size from two to six base pairs (2–6 bp). Different alleles are identified by the number of repeats.

silent mutation   A point mutation in coding DNA that results in the same amino acid. For example, if the DNA sequence that codes for the amino acid glycine changes from CCA to CCG, the same amino acid is produced, such that the mutation has no impact and is "silent."

single-nucleotide polymorphisms (SNPs)   Genetic markers defined by different bases (A, C, G, or T) at any given position in a DNA sequence.

site frequency spectrum   A method used to detect natural selection that looks at the relative occurrence of loci that have different allele frequencies.

stabilizing selection   Selection on a quantitative trait that selects against extreme phenotypic values (too small or too large) and selects for the intermediate values. Under this type of selection, the average phenotypic value is stabilized over time.

transition   A specific type of point mutation. The four nucleotide bases include two that are purines (A and G), that have two carbon–nitrogen rings, and two pyrimidines (C and T), that have one carbon–nitrogen ring. Transitions are point mutations between one purine and the other (A to G, or G to A) or between one pyrimidine and the other (C to T, or T to C).

translocation   A mutation where an entire section of a chromosome moves to another chromosome. Sometimes the DNA sequences are exchanged between chromosomes, and sometimes they are not.

transversion   A specific type of point mutation. The four nucleotide bases include two that are purines (A and G), that have two carbon–nitrogen rings, and two pyrimidines (C and T), that have one carbon–nitrogen ring. Transversions are point mutations between a purine and a pyrimidine.

Y-Chromosome DNA   The DNA on the Y chromosome, which is one form of a sex chromosome, and the other form is the X chromosome. Females have two X chromosomes (XX) and males have an X chromosome and a Y chromosome (XY). Most of the Y chromosome does not recombine, and is passed along from father to son intact.

REFERENCES

Akey JM (2010) Constructing genomic maps of positive selection in humans: Where do we go from here? *Genome Research* **19**:711–722.

Alvarez G, Ceballos FC, Quinteiro C (2009) The role of inbreeding in the extinction of a European royal dynasty. *PLoS One* **4**(4):e5174 (doi:10.1371/journal.pone.0005174).

Ammermann AJ, Cavalli-Sforza LL (1984) *The Neolithic Transition and the Genetics of Populations in Europe.* Princeton: Princeton University Press.

Aoki K (2002) Sexual selection as a cause of human skin colour variation: Darwin's hypothesis revisited. *Annals of Human Biology* **29**:589–608.

Arias E (2007) United States life tables, 2004. *National Vital Statistics Report*, vol. 56, no. 9. Hyattsville, MD: National Center for Health Statistics (available at http://www.cdc.gov/nchs/data/nvsr/nvsr56/nvsr56_09.pdf).

Ayala FJ (1982) *Population and Evolutionary Genetics: A Primer.* Menlo Park, CA: Benjamin/Cummings.

Balter M (2005) Are humans still evolving? *Science* **309**:234–237.

Bamshad MJ, Mummidi S, Gonzalez E, Ahuja SS, Dunn DM, Watkins WS, Wooding S, Stone AC, Jorde LB, Weiss RB, Ahuja SK (2002) A strong signature of balancing selection in the 5′ *cis*-regulatory region of *CCR5*. *Proceedings of the National Academy of Sciences USA* **99**:10539–10544.

Beall CM (2007) Two routes to functional adaptation: Tibetan and Andean high-altitude natives. *Proceedings of the National Academy of Sciences USA* **104**:8655–8660.

Beall CM, Cavalleri GL, Deng L, Elston RC, Gao Y, Knight J, Li C, Li JC, Liang Y, McCormack M, Montgomery HE, Pan H, Robbins PA, Shianna KV, Tam SC, Tsering N, Veeramah KR, Wang W, Wangdui P, Weale ME, Xu Z, Yang L, Zaman MJ, Zeng C, Zhang L, Zhang X, Zhaxi P, Zheng YT (2010) Natural selection on *EPAS1* (HIFα) associated with low hemoglobin concentration in Tibetan highlanders. *Proceedings of the National Academy of Sciences USA* **107**:11459–11464.

Beall CM, Steegmann Jr AT (2000) Human adaptation to climate: Temperature, ultraviolet radiation, and altitude. In *Human Biology: An Evolutionary and Biocultural Perspective*, Stinson S, Bogin B, Huss-Ashmore R, O'Rourke D, eds. New York: Wiley-Liss, pp. 163–224.

Beja-Pereira A, Luikart G, England PR, Bradley DG, Jann OC, Bertorelle G, Chamberlain AT, Nunes TP, Metodiev S, Ferrand N, Erhardt G (2003) Gene–culture coevolution between cattle milk protein genes and human lactase genes. *Nature Genetics* **35**:311–313.

Human Population Genetics, First Edition. John H. Relethford.
© 2012 Wiley-Blackwell. Published 2012 by John Wiley & Sons, Inc.

Bittles AH (2001) Consanguinity and its relevance to clinical genetics. *Clinical Genetics* **60**:89–98.

Bittles AH (2004) Genetic aspects of inbreeding and incest. In *Inbreeding, Incest, and the Incest Taboo*, Wolf AP, Durham WH, eds. Stanford: Stanford University Press, pp. 38–60.

Bittles AH, Black ML (2010) Consanguinity, human evolution, and complex diseases. *Proceedings of the National Academy of Sciences USA* **107**:1779–1786.

Bodmer WF, Cavalli-Sforza LL (1968) A migration matrix model for the study of random genetic drift. *Genetics* **59**:565–592.

Bodmer WF, Cavalli-Sforza LL (1976) *Genetics, Evolution, and Man*. San Francisco: Freeman.

Boyd WC (1950) *Genetics and the Races of Man*. Boston: Little, Brown.

Bramble DM, Lieberman DE (2004) Endurance running and the evolution of *Homo*. *Nature* **432**:345–352.

Brennan ER, Relethford JH (1983) Temporal variation in the mating structure of Sanday, Orkney Islands. *Annals of Human Biology* **10**:265–280.

Bustamante CD, Henn BM (2010) Shadows of early migration. *Nature* **468**:1044–1045.

Capelli C, Wilson JF, Richards M, Stumpf MPH, Gratix F, Oppenheimer S, Underhill P, Pascali VL, Ko T-M, Goldstein DB (2001) A predominately indigenous paternal heritage for the Austronesian-speaking peoples of insular Southeast Asia and Oceania. *American Journal of Human Genetics* **68**:432–443.

Cavalli-Sforza LL (1969) Genetic drift in an Italian population. *Scientific American* **221**(2): 30–37.

Cavalli-Sforza LL, Bodmer WF (1971) *The Genetics of Human Populations*. San Francisco: Freeman.

Cavalli-Sforza LL, Menozzi P, Piazza A (1993) Demic expansions and human evolution. *Science* **259**:639–646.

Cavalli-Sforza LL, Menozzi P, Piazza A (1994) *The History and Geography of Human Genes*. Princeton: Princeton University Press.

Cavalli-Sforza LL. Moroni A, Zei G (2004) *Consanguinity, Inbreeding, and Genetic Drift in Italy*. Princeton: Princeton University Press.

Chakraborty R (1986) Gene admixture in human populations: Models and predictions. *Yearbook of Physical Anthropology* **29**:1–43.

Chaplin G, Jablonski NG (2009) Vitamin D and the evolution of human skin depigmentation. *American Journal of Physical Anthropology* **139**:451–461.

Chaudhuri A, Polyakova J, Zbrzezna V, Williams K, Gulati S, Pogo AO (1993) Cloning of glycoprotein D cDNA, which encodes the major subunit of the Duffy blood group system and the receptor for the *Plasmodium vivax* malaria parasite. *Proceedings of the National Academy of Sciences USA* **90**:10793–10797.

Connell KH (1950) *The Population of Ireland 1750–1845*. Oxford: Clarendon Press.

Cook LM, Dennis RLH, Mani GS (1999) Melanic morph frequency in the peppered moth in the Manchester area. *Proceedings of the Royal Society of London B* **266**:293–297.

Coop G, Pickrell JK, Novembre J, Kudaravalli S, Li J, Absher D, Myers RM, Cavalli-Sforza LL, Feldman MW, Pritchard JK (2009) The role of geography in human adaptation. *PLoS Genetics* **5**(6):e1000500. doi:10.1371/journal/pgen.1000500.

Crawford MH (1973) The use of genetic markers of the blood in the study of the evolution of human populations. In *Methods and Theories of Anthropological Genetics*, Crawford MH, Workman PL, eds. Albuquerque: University of New Mexico Press, pp. 19–38.

Crawford MH, ed. (1984) *Current Developments in Anthropological Genetics*, vol. 3, *Black Caribs: A Case Study in Biocultural Adaptation*. New York: Plenum Press.

Crawford MH (1998) *The Origins of Native Americans: Evidence from Anthropological Genetics.* Cambridge: Cambridge University Press.

Crawford MN, ed (2007) *Anthropological Genetics: Theory, Methods and Applications.* Cambridge: Cambridge University Press.

Crawford MH, Gmelch G (1974) Human biology of the Irish Tinkers: Demography, ethnohistory, and genetics. *Social Biology* **21**:321–331.

Crawford MH, Mielke JH, eds. (1982) *Current Developments in Anthropological Genetics,* Vol. 2, *Ecology and Population Structure.* New York: Plenum Press.

Crawford MH, Workman PL, eds (1973) *Methods and Theories of Anthropological Genetics.* Albuquerque: University of New Mexico Press.

Croke DT, Tighe O, O'Neill C, Mayne PD (2000) The "Travellers": An isolate within the Irish population. In *Archaeogenetics: DNA and the Population Prehistory of Europe*, Renfrew C, Boyle K, eds. Cambridge: McDonald Institute for Archaeological Research, University of Cambridge, pp. 209–212.

Crow JF, Kimura M (1970) *An Introduction to Population Genetics Theory.* Minneapolis: Burgess.

Crow JH (1980) The estimation of inbreeding from isonymy. *Human Biology* **52**:1–14.

Crow JH, Mange AP (1965) Measurement of inbreeding from the frequency of marriages between persons of the same surname. *Eugenics Quarterly* **12**:199–203.

Dalton R (2003) The coast road. *Nature* **422**:10–12.

Derenko MV, Grzybowski T, Malyarchuk BA, Czarny J, Miscicka-Sliwka D, Zakharov IA (2001) The presence of mitochondrial haplogroup X in Altaians from South Siberia. *American Journal of Human Genetics* **69**:237–241.

Devor EJ, Crawford MH (1980) Population structure and admixture in transplanted Tlax-caltecan populations. *American Journal of Physical Anthropology* **52**:485–490.

Devor EJ, Crawford MH, Koertvelyessy T (1983) Marital structure and genetic heterogeneity of Ramea Island, Newfoundland. *American Journal of Physical Anthropology* **61**:401–409.

Diamond J (1988) Express train to Polynesia. *Nature* **336**:307–308.

Diamond J (1997) *Guns, Germs, and Steel.* New York: Norton.

Elias PM, Menon G, Wetzel BK, Williams JW (2010) Barrier requirements as the evolutionary "driver" of epidermal pigmentation in humans. *American Journal of Human Biology* **22**:526–537.

Eller E, Hawks J, Relethford JH (2004) Local extinction and recolonization, species effective population size, and modern human origins. *Human Biology* **76**:689–709.

Elseth GD, Baumgardner KD (1981) *Population Biology.* New York: Van Nostrand.

Falconer DS, TFC Mackay (1996) *Introduction to Quantitative Genetics*, 4 edition. Upper Saddle River, NJ: Benjamin-Cummings.

Falush D, Stephens M, Pritchard JK (2003) Inference of population structure using multilocus genotype data: Linked loci and correlated allele frequencies. *Genetics* **164**:1567–1587.

Fix AG (1978) The role of kin-structured migration in genetic microdifferentiation. *Annals of Human Genetics* **41**:329–339.

Fix AG (1999) *Migration and Colonization in Human Microevolution.* Cambridge: Cambridge University Press.

Foster EA, Jobling MA, Taylor PG, Donnelly P, de Knijffs P, Mieremet R, Zerjal T, Tyler-Smith C (1998) Jefferson fathered slave's last child. *Nature* **396**:27–28.

Friedlaender JS (1975) *Patterns of Human Variation: The Demography, Genetics, and Phenetics of Bougainville Islanders.* Cambridge, MA: Harvard University Press.

Friedlaender JS, Friedlaender FR, Reed FA, Kidd KK, Kidd JR, Chambers GK, Lea RA, Loo J-H, Koki G, Hodgson JA, Merriwether DA, Weber JL (2008) The genetic structure of Pacific Islanders. *PLoS Genetics* **4**(1):e19 (doi:10.1371/journal.pgen.0040019).

Frisancho AR (1993) *Human Adaptation and Accommodation*. Ann Arbor: University of Michigan Press.

Frisancho AR, Baker PT (1970) Altitude and growth: A study of the patterns of physical growth of a high altitude Peruvian Quechua population. *American Journal of Physical Anthropology* **32**:279–292.

Gagneux P, Wills C, Gerloff U, Tautz D, Morin A, Boesch C, Fruth B, Hohmann G, Ryder A, Woodruff DS (1999) Mitochondrial sequences show diverse evolutionary histories of African hominoids. *Proceedings of the National Academy of Science USA* **96**:5077–5082.

Galvani AP, Slatkin M (2003) Evaluating plague and smallpox as historical selective pressures for the *CCR5-Δ32* HIV-resistance allele. *Proceedings of the National Academy of Sciences USA* **100**:15276–15279.

Garrigan D, Hammer MF (2006) Reconstructing human origins in the genomic era. *Nature Reviews Genetics* **7**:669–680.

Gibbons A (2010) Tracing evolution's recent fingerprints. *Science* **329**:740–742.

Giles E, Hansen AT, McCullough JM, Metzger DG, Wolpoff MH (1968) Hydrogen cyanide and phenylthiocarbamide sensitivity, mid-phalangeal hair and color blindness in Yucatán, Mexico. *American Journal of Physical Anthropology* **28**:203–212.

Gillespie JH (2004) *Population Genetics: A Concise Guide*. Second edition. Baltimore: Johns Hopkins University Press.

Glazko GV, Nei M (2003) Estimation of divergence times for major lineages of primate species. *Molecular Biology and Evolution* **20**:423–434.

Gmelch G (1977) *The Irish Tinkers: The Urbanization of an Itinerant People*. Menlo Park, CA: Cummings.

Goebel T, Waters MR, O'Rourke DH (2008) The Late Pleistocene dispersal of modern humans in the Americas. *Science* **319**:1497–1502.

Gómez-Pérez L, Alfonso-Sánchez MA, Dipierri JE, Alfaro E, García-Obregón S, De Pancarbo MM, Bailliet G, Peńa JA (2011) Microevolutionary processes due to landscape features in the Province of Jujuy (Argentina). *American Journal of Human Biology* **23**:177–184.

Goodman M, Cronin JE (1982) Molecular anthropology: Its development and current directions. In *A History of American Physical Anthropology 1930–1980*, Spencer F, ed. New York: Academic Press, pp. 105–146.

Gould SJ (1999) *Rock of Ages: Science and Religion in the Fullness of Life*. New York: Ballantine.

Green RE, Krause J, Briggs AW, Maricic T, Stenzel U, Kircher M, Patterson N, Li H, Zhai W, Fritz MH-Y, Hansen NF, Durand EY, Malaspinas A-S, Jensen JD, Marques-Bonet T, Alkan C, Prüfer K, Meyer M, Burbano HA, Good JM, Schultz R, Aximu-Petri A, Butthof A, Höber B, Höffner B, Siegemund M, Weihmann A, Nusbaum C, Lander ES, Russ C, Novod N, Affourtit J, Egholm M, Verna C, Rudan P, Brajkovic D, Kucan Z, Gušic I, Doronichev VB, Golovanova LV, Lalueza-Fox C, de la Rasilla M, Fortea J, Rosas A, Schmitz RW, Johnson PLF, Eichler EE, Falush D, Birney E, Mullikin JC, Slatkin M, Nielsen R, Kelso J, Lachmann M, Reich D, Pääbo S (2010) A draft sequence of the Neandertal genome. *Science* **328**:710–722.

Green RE, Malaspinas, A-S, Krause J, Briggs AW, Johnson PLF, Uhler C, Meyer M, Good JM, Maricic T, Stenzel U, Prüfer K, Siebauer M, Burbano HA, Ronan M, Rothberg JM, Egholm M, Rudan P, Brajković D, Kućan Z, Gušic I, Wikström M, Laakkonen L, Kelso J, Slatkin M, Pääbo S (2008) A complete Neanderthal mitochondrial genome sequence determined by high-throughput sequencing. *Cell* **134**:416–426.

Greksa LP (1996) Evidence for a genetic basis to the enhanced total lung capacity of Andean highlanders. *Human Biology* **68**:119–129.

Haak W, Balanovsky O, Sanchez JJ, Koshel S, Zaporozhchenko V, Adler CJ, De Sarkissian CSI, Brandt G, Schwarz C, Nicklisch N, Dresely V, Fritsch B, Balanovska E, Villems R, Meller H, Alt KW, Cooper A, Genographics Consortium (2010) Ancient DNA from European Early Neolithic farmers reveals their Near Eastern affinities. *PLoS Biology* **8**(11):e1000536 (doi:10.1371/journal/pbio.1000536).

Hamblin MT, Di Rienzo A (2000) Detection of the signature of natural selection in humans: Evidence from the Duffy blood group locus. *American Journal of Human Genetics* **66**:1669–1679.

Hamblin MT, Thompson EE, Di Rienzo A (2002) Complex signatures of natural selection at the Duffy blood group locus. *American Journal of Human Genetics* **70**:369–383.

Hamilton MB (2009) *Population Genetics*. Hoboken, NJ: Wiley-Blackwell.

Hanis CL, Chakraborty R, Ferrell RE, Schull WJ (1986) Individual admixture estimates: Disease associations and individual risk of diabetes and gallbladder disease among Mexican-Americans in Starr County, Texas. *American Journal of Physical Anthropology* **70**:433–441.

Harding RM, Healy E, Ray AJ, Eliis NS, Flanagan N, Todd C, Dixon C, Sajantila A, Jackson IJ, Birch-Machin MA, Rees JL (2000) Evidence for variable selective pressures at MC1R. *American Journal of Human Genetics* **66**:1351–1361.

Hardy GH (1908) Mendelian proportions in a mixed population. *Science* **28**:49–50.

Harpending H, Jenkins T (1973) Genetic distance among Southern African populations. In *Methods and Theories of Anthropological Genetics*, ed. by MH Crawford MH, Workman PL, eds. Albuquerque: University of New Mexico Press, pp. 177–200.

Hartl DL, Clark AG (2007) *Principles of Population Genetics*. 4th edition. Sunderland, MA: Sinauer.

Hawks H (2009) Update to Eller et al.'s "Local extinction and recolonization, species effective size, and modern human origins (2004)." *Human Biology* **81**:825–828.

Hawks J, Wang ET, Cochran GM, Harpending HC, Moyziz RK (2007) Recent acceleration of human adaptive evolution. *Proceedings of the National Academy of Sciences USA* **104**:20753–20758.

Hedrick PW (2005) *Genetics of Populations*, 3rd edition. Sudbury, MA: Jones & Bartlett.

Hodgson JA, Disotell TR (2008) No evidence of a Neandertal contribution to modern human genetic diversity. *Genome Biology* **9**:206 (doi:10.1186/gb-2008-9-2-206).

Howell N (2000) *Demography of the Dobe !Kung*, 2nd edition. New York: Aldine de Gruyter.

Hummel S, Schmidt D, Kremeyer B, Herrmann B, Oppermann M (2005) Detection of the *CCR5-Δ32* HIV resistance gene in Bronze Age skeletons. *Genes and Immunity* **6**:371–374.

International HapMap Consortium (2007) A second generation human haplotype map of over 3.1 million SNPs. *Nature* **449**:851–861.

Jablonski NG, Chaplin G (2000) The evolution of human skin coloration. *Journal of Human Evolution* **39**:57–106.

Jacquard A (1974) *The Genetic Structure of Populations*. New York: Springer-Verlag.

Jobling MA, Hurles ME, Tyler-Smith C (2004) *Human Evolutionary Genetics: Origins, Peoples & Disease*. New York: Garland.

Jorde LB (1980) The genetic structure of subdivided human populations: A review. In *Current Developments in Anthropological Genetics*, vol. 1, *Theory and Methods*, Mielke JH, Crawford MH, eds. New York: Plenum Press, pp. 135–208.

Jorde LB, Workman PL, Eriksson AW (1982) Genetic microevolution in the Åland Islands, Finland. In *Current Developments in Anthropological Genetics*, vol. 2, *Ecology and Population Structure*, Crawford MH, Mielke JH, eds. New York: Plenum Press, pp. 333–366.

Karafet TM, Mendez FL, Meilerman MB, Underhill PA, Zegura SL, Hammer MF (2008) New binary polymorphisms reshape and increase resolution of the human y chromosomal haplogroup tree. *Genome Research* **18**:830–838.

Karn MN, Penrose LS (1952) Birth weight and gestation time in relation to maternal age, parity and infant survival. *Annals of Eugenics* **16**:147–164.

Kayser M, Brauer S, Weiss G, Underhill PA, Roewer L, Schiefenhövel W, Stoneking M (2000) Melanesian origin of Polynesian Y chromosomes. *Current Biology* **10**:1237–1246.

Kennedy RE Jr (1973) *The Irish: Emigration, Marriage, and Fertility*. Berkeley: University of California Press.

Kimura M, Ohta T (1969) The average number of generations until fixation of a mutant gene in a finite population. *Genetics* **61**:763–771.

Konigsberg LW (2000) Quantitative variation and genetics. In *Human Biology: An Evolutionary and Biocultural Perspective*, Stinson S, Bogin B, Huss-Ashmore R, O'Rourke D, eds. New York: Wiley-Liss, pp. 135–162.

Krings M, Stone A, Schmitz RW, Krainitzki H, Stoneking M, Pääbo S (1997) Neandertal DNA sequences and the origin of modern humans. *Cell* **90**:19–30.

Lalueza-Fox C, Römpler H, Caramelli D, Stäubert C, Catalano G, Hughes D, Rohland N, Pilli E, Longo L, Condemi S, de la Rasilla M, Fortea J, Rossa A, Stoneking M, Schöneberg T, Betranpetit J, Hofreiter M (2007) A melanocortin 1 receptor allele suggests varying pigmentation among Neanderthals. *Science* **318**:1453–1455.

Langaney A, Gomila J (1973) Bedik and Niokholonko intra and inter-ethnic migration. *Human Biology* **45**:137–150.

Larsen CS (2000) *Skeletons in Our Closet: Revealing Our Past through Bioarchaeology*. Princeton: Princeton University Press.

Lasker GW (1985) *Surnames and Genetic Structure*. Cambridge: Cambridge University Press.

Leonard WR (2000) Human nutritional evolution. In *Human Biology: An Evolutionary and Biocultural Perspective*, Stinson S, Bogin B, Huss-Ashmore R, O'Rourke D, eds. New York: Wiley-Liss, pp. 295–343.

Leslie PW (1985) Potential mates analysis and the study of human population structure. *Yearbook of Physical Anthropology* **28**:53–78.

Livingstone FB (1958) Anthropological implications of sickle cell gene distribution in West Africa. *American Anthropologist* **60**:533–562.

Livingstone FB (1967) *Abnormal Hemoglobins in Human Populations: A Summary and Interpretation*. Chicago: Aldine.

Livingstone FB (1984) The Duffy blood groups, vivax malaria, and malaria selection in human populations: A review. *Human Biology* **56**:413–425.

Long JC (1986) The allelic correlation structure of Gainj- and Kalam-speaking people. I. The estimation and interpretation of Wright's F-statistics. *Genetics* **112**:629–647.

Long JC (1991) The genetic structure of admixed populations. *Genetics* **127**:417–428.

Long JC, Smouse PE, Wood JW (1987) The allelic correlation structure of Gainj- and Kalam-speaking people. II. The genetic distance between population subdivisions. *Genetics* **117**:273–283.

Madrigal L, Barbujani G (2007) Partitioning of genetic variation in human populations and the concept of race. In *Anthropological Genetics: Theory, Methods and Applications*, Crawford MH, ed. Cambridge: Cambridge University Press, pp. 19–37.

Madrigal L, Kelly W (2007) Human skin-color sexual dimorphism: A test of the sexual selection hypothesis. *American Journal of Physical Anthropology* **132**:470–482.

Marks J (1994) Black, White, other. *Natural History* **103**(12): 32–35.

Marks J (2011) *The Alternative Introduction to Biological Anthropology*. New York: Oxford University Press.

Marks J, Lyles RB (1994) Rethinking genes. *Evolutionary Anthropology* **3**:139–146.

Mayr E (2000) Darwin's influence on modern thought. *Scientific American* **283**(1): 78–83.

McDougall I, Brown FH, Fleagle JG (2005) Stratigraphic placement and age of modern humans from Kibish, Ethiopia. *Nature* **433**:733–736.

Menozzi P, Piazza A, Cavalli-Sforza LL (1978) Synthetic maps of human gene frequencies in Europe. *Science* **201**:786–792.

Mielke JH, Crawford MH, eds. (1980) *Current Developments in Anthropological Genetics*, vol. 1, *Theory and Methods*. New York: Plenum Press.

Mielke JH, Devor EJ, Kramer PL, Workman PL, Eriksson AW (1982) Historical population structure of the Åland Islands, Finland. In *Current Developments in Anthropological Genetics*, vol. 2, *Ecology and Population Structure*, Crawford MH, Mielke JH, eds. New York: Plenum Press, pp. 255–332.

Mielke JH, Konigsberg LW, Relethford JH (2011) *Human Biological Variation*, 2nd edition. New York: Oxford University Press.

Mielke JH, Relethford JH, Eriksson AW (1994) Temporal trends in migration in the Åland Islands: Effects of population size and geographic distance. *Human Biology* **66**:399–410.

Mielke JH, Workman PL, Fellman J, Eriksson AW (1976) Population structure of the Åland Islands, Finland. *Advances in Human Genetics* **6**:241–321.

Morton N, ed. (1973) *Genetic Structure of Populations*. Honolulu: University Press of Hawaii.

Murphy M, McHugh B, Tighe O, Mayne P, O'Neill C, Naughten E, Croke DT (1999) Genetic basis of transferase-deficient galactosaemia in Ireland and the population history of the Irish Travellers. *European Journal of Human Genetics* **7**:549–554.

Natarajan V (2008) What Einstein meant when he said "God does not play dice. ..." *Resonance* **13**:655–661.

Nei M (1972) Genetic distance between populations. *American Naturalist* **106**:283–292.

Nei M (1987) *Molecular Evolutionary Genetics*. New York: Columbia University Press.

Nei M, Kumar S (2000) *Molecular Evolution and Phylogenetics*. Oxford: Oxford University Press.

Nemecek S (2000) Who were the first Americans? *Scientific American* **283**(3): 80–88.

North KE, Martin LJ, Crawford MH (2000) The origins of the Irish travelers and the genetic structure of Ireland. *Annals of Human Biology* **27**:453–465.

Norton HL, Kittles RA, Parra E, McKeigue P, Mao X, Cheng K, Canfield VA, Bradley DG, McEvoy B, Shriver MD (2007) Genetic evidence for the convergent evolution of light skin in Europeans and East Asians. *Molecular Biology and Evolution* **24**:710–722.

Ohta T (1992) The nearly neutral theory of molecular evolution. *Annual Review of Ecology and Systematics* **23**:263–286.

O'Rourke DH (2007) Ancient DNA and its application to the reconstruction of human evolution and history. In *Anthropological Genetics: Theory, Methods and Applications*, Crawford MH, ed. Cambridge: Cambridge University Press, pp. 210–231.

Parra EJ, Kittles RA, Argyropoulos G, Pfaff CL, Hiester K, Bonila C, Sylvester N, Parrish-Gause D, Garvey WT, Jin L, McKeigue PM, Kamboh MI, Ferrell RE, Pollitzer WS, Shriver MD (2001) Ancestral proportions and admixture dynamics in geographically defined

African Americans living in South Carolina. *American Journal of Physical Anthropology* **114**:18–29.

Parra EJ, Marchini A, Akey J, Martinson J, Batzer MA, Cooper R, Forrester T, Allison DB, Deka R, Ferrell RE, Shriver MD (1998) Estimating African American admixture proportions by use of population-specific alleles. *American Journal of Human Genetics* **63**:1839–1851.

Pinhasi R, von Cramon-Taubadel N (2009) Craniometric data supports demic diffusion model for the spread of agriculture into Europe. *PLoS One* **4**(8):e6747 (`doi:10.1371/journal/pone.0006747`).

Population Reference Bureau (2010) *2010 World Population Data Sheet*. Washington, DC: Population Reference Bureau (available at `http://www.prb.org/pdf10/10wpds_eng.pdf`).

Post PW, Daniels F Jr, Binford RT Jr (1975) Cold injury and the evolution of "white" skin. *Human Biology* **47**:65–80.

Pritchard JK, Stephens M, Donnelly P (2000) Inference of population structure using multilocus genotype data. *Genetics* **155**:945–959.

Provine WB (1971) *The Origins of Theoretical Population Genetics*. Chicago: University of Chicago Press.

Ramachandran S, Deshpande O, Roseman CC, Rosenberg NA, Feldman MW, Cavalli-Sforza LL (2005) Support for the relationship of genetic and geographic distance in human populations for a serial founder effect originating in Africa. *Proceedings of the National Academy of Sciences* **102**:15942–15947.

Redd AJ, Takezaki N, Sherry ST, McGarvey ST, Sofro ASM, Stoneking M (1995) Evolutionary history of the COII/tRNA^{Lys} intergenic 9 base pair deletion in human mitochondrial DNAs from the Pacific. *Molecular Biology and Evolution* **12**:604–615.

Reed FA, Aquardo CF (2006) Mutation, selection and the future of human evolution. *Trends in Genetics* **22**:479–484.

Reed TE (1969) Caucasian genes in American Negroes. *Science* 165:762–768.

Reich D, Green RE, Kircher M, Krause J, Patterson N, Durand EY, Viola B, Briggs AW, Stenzel U, Johnson PLF, Maricic T, Good JM, Marques-Bonet T, Alkan C, Fu Q, Mallick S, Li H, Meyer M, Eichler EE, Stoneking M, Richards M, Talamo S, Shunkov MV, Derevianko AP, Hublin J-J, Kelso J, Slatkin M, Pääbo S (2010) Genetic history of an archaic hominin group from Denisova Cave in Siberia. *Nature* **468**:1053–1060.

Reid RM (1973) Inbreeding in human populations. In *Methods and Theories of Anthropological Genetics*, Crawford MH and Workman PL eds. Albuquerque: University of New Mexico Press, pp. 83–116.

Relethford JH (1981) The effect of consanguinity avoidance on potential inbreeding in captive nonhuman primate colonies. *American Journal of Primatology* **1**:336 (abstract).

Relethford JH (1986) Density-dependent migration and human population structure in historical Massachusetts. *American Journal of Physical Anthropology* **69**:377–388.

Relethford JH (1988) Estimation of kinship and genetic distance from surnames. *Human Biology* **60**:475–492.

Relethford JH (1991) Effect of population size on marital migration distance. *Human Biology* **63**:629–641.

Relethford JH (1992) Cross-cultural analysis of migration rates: Effects of geographic distance and population size. *American Journal of Physical Anthropology* **89**:459–466.

Relethford JH (1997) Hemispheric difference in human skin color. *American Journal of Physical Anthropology* **104**:449–457.

Relethford JH (2000) Human skin color diversity is highest in Sub-Saharan African populations. *Human Biology* **72**:773–780.

Relethford JH (2001) *Genetics and the Search for Modern Human Origins.* New York: Wiley-Liss.

Relethford JH (2003) *Reflections of Our Past: How Human History is Revealed in Our Genes.* Boulder: Westview Press.

Relethford JH (2004) Global patterns of isolation by distance based on genetic and morphological data. *Human Biology* **76**:499–513.

Relethford JH (2007) The use of quantitative traits in anthropological genetic studies of population structure and history. In *Anthropological Genetics: Theory, Methods and Applications*, Crawford MH, ed. Cambridge: Cambridge University Press, pp. 187–209.

Relethford JH (2008a) Genetic evidence and the modern human origins debate. *Heredity* **100**:555–563.

Relethford JH (2008b) Geostatistics and spatial analysis in biological anthropology. *American Journal of Physical Anthropology* **136**:1–10.

Relethford JH (2009) Race and global patterns of phenotypic variation. *American Journal of Physical Anthropology* **139**:16–22.

Relethford JH (2010) Population-specific deviations of global craniometric variation from a neutral model. *American Journal of Physical Anthropology* **142**:105–111.

Relethford JH (2011) Understanding human cranial variation in light of modern human origins. In *Origins of Modern Humans*, Smith F, Ahern J, eds. Hoboken, NJ: Wiley-Blackwell (in press).

Relethford JH, Blangero J (1990) Detection of differential gene flow from patterns of quantitative variation. *Human Biology* **62**:5–25.

Relethford JH, Crawford MH (1995) Anthropometric variation and the population history of Ireland. *American Journal of Physical Anthropology* **96**:25–38.

Relethford JH, Crawford MH, Blangero J (1997) Genetic drift and gene flow in post-Famine Ireland. *Human Biology* **69**:443–465.

Relethford JH, Jaquish CE (1988) Isonymy, inbreeding, and demographic variation in historical Massachusetts. *American Journal of Physical Anthropology* **77**:243–252.

Relethford JH, Jorde LB (1999) Genetic evidence for larger African population size during recent human evolution. *American Journal of Physical Anthropology* **108**:251–260.

Relethford JH, Mielke JH (1994) Marital exogamy in the Åland Islands, 1750–1949. *Annals of Human Biology* **21**:13–21.

Richards M (2003) The Neolithic invasion of Europe. *Annual Review of Anthropology* **32**:135–162.

Roberts DF (1968) Genetic effects of population size reduction. *Nature* **220**:1084–1088.

Robins AH (1991) *Biological Perspectives on Human Pigmentation.* Cambridge: Cambridge University Press.

Robins AH (2009) The evolution of light skin color: Role of vitamin D disputed. *American Journal of Physical Anthropology* **139**:447–450.

Rogers A (1991) Doubts about isonymy. *Human Biology* **63**:663–668.

Rogers AR, Harpending HC (1986) Migration and genetic drift in human populations. *Evolution* **40**:1312–1327.

Rogers AR, Harpending H (1992) Population growth makes waves in the distribution of pairwise genetic differences. *Molecular Biology and Evolution* **9**:552–569.

Rogers AR, Iltis D, Wooding S (2004) Genetic variation at the MC1R locus and the time since loss of human body hair. *Current Anthropology* **45**:105–108.

Rogers L (1987) Concordance in isonymy and pedigree measures of inbreeding: The effects of sample composition. *Human Biology* **59**:753–767.

Rohlf FJ, Sokal RR (1995) *Statistical Tables*. Third edition. New York: Freeman.

Rosenberg NA, Li LM, Ward R, Pritchard JK (2003a) Informativeness of genetic markers for inference of ancestry. *American Journal of Human Genetics* **73**:1402–1422.

Rosenberg NA, Pritchard JK, Weber JL, Cann HM, Kidd KK, Zhivotovsky LA, Feldman MW (2003b) Response to comment on "Genetic structure of human populations." *Science* **300**:1877c.

Roychoudhury AK, Nei M (1988) *Human Polymorphic Genes: World Distribution*. New York: Oxford University Press.

Sachs J, Malaney P (2002) The economic and social burden of malaria. *Nature* **415**:680–685.

Salzano FM, Neel JV, Maybury-Lewis D (1967) Further studies on the Xavante Indians I. Demographic data on two additional villages: Genetic structure of the tribe. *American Journal of Human Genetics* **19**:463–489.

Salzano FM, Weimer TA, Franco MHLP, Mestriner MA, Simões AL, Constans J, De Melo e Freitas MJ (1985) Population structure and blood genetics of the Pacaás Novos Indians of Brazil. *Annals of Human Biology* **12**:241–249.

Sanger R, Tippett P, Gavin J (1971) The X-linked blood group system Xg: Tests on unrelated people and families of Northern European ancestry. *Journal of Medical Genetics* **8**:427–433.

Sarich VM (1971) A molecular approach to the question of human origins. In *Background for Man: Readings in Physical Anthropology*, Dohlinow P, Sarich VM, eds. Boston: Little, Brown, pp. 60–81.

Sarich V, Wilson A (1967) Immunological time scale for hominoid evolution. *Science* **158**:1200–1203.

Schull WJ (1972) Genetic implications of population breeding structure. In *The Structure of Human Populations*, Harrison GA, Boyce AJ, eds. Oxford: Clarendon Press, pp. 146–164.

Seixas S, Ferrand N, Rocha J (2002) Microsatellite variation and evolution of the human Duffy blood group polymorphism. *Molecular Biology and Evolution* **19**:1802–1806.

Seldin MF, Price AL (2008) Application of ancestry informative markers to association studies in European Americans. *PLoS Genetics* **4**(1):e5 (doi:10,1371/journal.pgen.0040005).

Simons AM (2002) The continuity of microevolution and macroevolution. *Journal of Evolutionary Biology* **15**:688–701.

Simonson TS, Tang Y, Huff CD, Yun H, Qin G, Witherspoon DJ, Bai Z, Lorenzo FR, Xing J, Jorde LB, Prchal JT, Ge R (2010) Genetic evidence for high-altitude adaptation in Tibet. *Science* **329**:72–75.

Smith CAB (1969) Local fluctuations in gene frequencies. *Annals of Human Genetics* **32**:251–260.

Smith FH, Janković I, Karavanić I (2005) The assimilation model, modern human origins in Europe, and the extinction of the Neandertals. *Quaternary International* **137**:7–19.

Smouse PE, Long JC (1992) Matrix correlation analysis in anthropology and genetics. *Yearbook of Physical Anthropology* **35**:187–213.

Spuhler JN (1989) Update to Spuhler and Kluckhohn's "Inbreeding coefficients of the Ramah Navaho population." *Human Biology* **61**:726–730.

Spuhler JN, Kluckhohn C (1953) Inbreeding coefficients of the Ramah Navaho population. *Human Biology* **25**:295–317.

Stephens JC, Reich DE, Goldstein DB, Shin HD, Smith MW, Carrington M, Winkler C, Huttley GA, Allikmets R, Schriml L, Gerrard B, Malasky M, Ramos MD, Morlot S,

Tzetis M, Hanson M, Kalaydjieva L, Glavac D, Gasparini P, Kanavakis E, Claustres M, Kambouris M, Ostrer H, Duff G, Baranov V, Sibul H, Metspalu, Goldman D, Martin N, Duffy D, Schmidtke J, Estivill X, O'Brien SJ, Dean M (1998) Dating the origin of the *CCR5-Δ32* AIDS-resistance allele by the coalescence of haplotypes. *American Journal of Human Genetics* **62**:1507–1515.

Stern C (1943) The Hardy-Weinberg law. *Science* **97**:137–138.

Stern C (1965) Mendel and human genetics. *Proceedings of the American Philosophical Society* **109**:216–226.

Stoneking M (1993) DNA and recent human evolution. *Evolutionary Anthropology* **2**:60–73.

Storfer A, Murphy MA, Evans JS, Goldberg CS, Robinson S, Spear SF, Dezzani R, Delmelle E, Vierling L, Waits LP (2007) Putting the "landscape" in landscape genetics. *Heredity* **98**:128–142.

Storz JF (2010) Genes for high altitudes. *Science* **329**:40–41.

Sykes B, Leiboff A, Low-Beer J, Tetzner S, Richards M (1995) The origins of the Polynesians: An interpretation from mitochondrial lineage analysis. *American Journal of Human Genetics* **57**:1463–1475.

Tanner JM (1990) *Foetus into Man: Physical Growth from Conception to Maturity*, revised and enlarged edition. Cambridge, MA: Harvard University Press.

Tattersall I (2008) *The World from Beginnings to 4000 BCE*. New York: Oxford University Press.

Templeton AR (2005) Haplotype trees and modern human origins. *Yearbook of Physical Anthropology* **48**:33–59.

Tishkoff SA, Gonder MK (2007) Human origins within and out of Africa. In *Anthropological Genetics: Theory, Methods and Applications*, Crawford MH, ed. Cambridge: Cambridge University Press, pp. 337–379.

Tishkoff SA, Reed FA, Ranciaro A, Voight BF, Babbitt CC, Silverman JS, Powell K, Mortensen HM, Hirbo JB, Osman M, Ibrahim M, Omar SA, Lema G, Nyambo TB, Ghori J, Bumpstead S, Pritchard JK, Wray GA, Deloukas P (2007) Convergent adaptation of human lactase persistence in Africa and Europe. *Nature Genetics* **39**:31–40.

Trinkaus E (2007) European early modern humans and the fate of the Neandertals. *Proceedings of the National Academy of Sciences USA* **104**:7367–7372.

Underhill PA, Passarino G, Lin AA, Marzuki S, Oefner PJ, Cavalli-Sforza LL, Chambers GK (2001) Maori origins, Y-chromosome haplotypes and implications for human history in the Pacific. *Human Mutation* **17**:271–280.

von Cramon-Taubadel N, Weaver TD (2009) Insights from a quantitative genetic approach to human morphological evolution. *Evolutionary Anthropology* **18**:237–240.

Wagner CL, Greer FR, Section on Breastfeeding and Committee on Nutrition (2008) Prevention of rickets and vitamin D deficiency in infants, children, and adolescents. *Pediatrics* **122**:1142–1152.

Wang ET, Kodama G, Baldi P, Moyzis (2006) Global landscape of recent inferred Darwinian selection for *Homo sapiens*. Proceedings of the National Academy of Sciences USA **103**:135–140.

Waters MR, Forman SL, Jennings TA, Nordt LC, Driese SG, Feinberg JM, Keene JL, Halligan J, Lindquist A, Pierson J, Hallmark CT, Collins MB, Wiederhold JE (2011) The Buttermilk Creek Complex and the origins of Clovis at the Debra L. Friedkin Site, Texas. *Science* **331**:1599–1603.

Weeks JR (2005) *Population: An Introduction to Concepts and Issues*, 9th edition. Belmont, CA: Wadsworth.

Weiner J (1994) *The Beak of the Finch: A Story of Evolution in Our Time*. New York: Vintage.

Weiss KM (1984) On the number of members of the genus *Homo* who have ever lived, and some evolutionary implications. *Human Biology* **56**:637–649.

Weiss KM, Buchanan AV (2009) *The Mermaid's Tale: Four BillionYears of Cooperation in the Making of Living Things*. Cambridge: Harvard University Press.

Weiss KM, Kurland JA (2007) Going on an antedate: A strange history of imperfect perfect proportions. *Evolutionary Anthropology* **16**:204–209.

White TD, Asfaw B, DeGusta D, Gilbert H, Richards GD, Suwa G, Howell FC (2003) Pleistocene *Homo sapiens* from Middle Awash, Ethiopia. *Nature* **423**:742–747.

Whiten A, Goodall J, McGrew WC, Nishida T, Reynolds V, Sugiyama Y, Tutin CEG, Wrangham RW, Boesch (1999) Culture in chimpanzees. *Nature* **399**:628–685.

Williams EM, Harper PS (1977) Genetic study of Welsh gypsies. *Journal of Medical Genetics* **14**:172–176.

Williams-Blangero S, Blangero J (1989) Anthropometric variation and the genetic structure of the Jirels of Nepal. *Human Biology* **61**:1–12.

Wolf CM, Myrianthopoulos NC (1973) Polydactyly in American Negroes and Whites. *American Journal of Human Genetics* **25**:397–404.

Wollstein A, Lao O, Becker C, Brauer S, Trent RJ, Nürnberg, Stoneking M, Kayser M (2010) Demographic history of Oceania inferred from genome-wide data. *Current Biology* **20**:1983–1992.

Wood B, Richmond BG (2000) Human evolution: Taxonomy and paleobiology. *Journal of Anatomy* **196**:19–60.

Wood JW, Smouse PJ, Long JC (1985) Sex-specific dispersal patterns in two human populations of highland New Guinea. *American Naturalist* **125**:747–768.

Workman PL, Harpending H, Lalouel JM, Lynch C, Niswander JD, Singleton R (1973) Population studies on Southwestern Indian tribes. VI. Papago population structure: A comparison of genetic and migration analyses. In *Genetic Structure of Populations*, Morton NE, ed. Honolulu: University of Press of Hawaii, pp. 166–194.

Workman PL, Niswander JD (1970) Population studies on southwestern Indian tribes. II. Local genetic differentiation in the Papago. *American Journal of Human Genetics* **22**:24–49.

Wright S (1969) *Evolution and the Genetics of Populations*, vol. 2, *The Theory of Gene Frequencies*. Chicago: University of Chicago Press.

Yi X, Liang Y, Huerta-Sanchez E, Jin X, Cuo ZXP, Pool JE, Xu X, Jiang H, Vinckenbosch N, Korneliussen TS, Zheng H, Liu T, He W, Li K, Luo R, Nie X, Wu H, Zhao M, Cao H, Zou J, Shan Y, Shuzheng L, Yang Q, Asan, Ni P, Tian G, Xu J, Liu X, Jiang T, Wu R, Zhou G, Tang M, Qin J, Wang Y, Feng S, Li G, Huasang, Luosang J, Wang W, Chen F, Wang Y, Zheng X, Li Z, Bianba Z, Yang G, Wang X, Tang S, Gao G, Chen Y, Luo Z, Gusang L, Cao Z, Zhang Q, Ouyang W, Ren X, Liang H, Zheng H, Huang Y, Li J, Bolund L, Kristiansen K, Li Y, Zhang Y, Zhang X, Li R, Li S, Yang H, Nielsen R, Wang J, Wang J (2010) Sequencing of 50 human exomes reveals adaptation to high altitude. *Science* **329**:75–78.

INDEX

Absolute fitness, 140, 184–185, 257
  in natural selection, 140–141
Accumulated admixture (*M*)
  per generation admixture *vs.*, 235–236
  in simple admixture model, 227–228
Adaptation, 203
  to disease and dietary change, 190
  evolutionary forces and, 20
  genetic drift and, 20
  among high-altitude populations,
    192–193
Adaptive value, of Duffy allele, 188–189
Admixture, 257
  to African-Americans, 253–255
  defined, 226
  estimating, 228–229
  extending analysis of, 229–230
  genetic drift *vs.* gene flow and, 231
  in human populations, 226–230
  in Native American origins, 249
  simple model of, 226–228
Admixture estimating model, 228–229
Advantage, in natural selection, 139
Aerobic capacity, in low- *vs.* high-altitude
    populations, 192
Africa
  in Duffy negative allele geography,
    187–189
  hemoglobin *S* allele in, 183–187
  hominin dispersal outside, 195–196
  human evolution in, 243–245, 256
  human skin color in, 194, 196
  lactase persistence allele in, 191–192
  in Native American origins, 248–249

African-Americans
  admixture to, 253–255
  European gene flow into, 226, 228, 231
African apes
  evolution of, 88–92, 96
  relationship to other apes and to humans,
    89, 91–92, 96
African replacement model, of human
    evolution, 243
Africans, Y chromosome haplogroups
    among, 95–97
Agriculture, 198–202
  genetics and spread of, 249–251
  in human population explosion, 20
  in spread of malaria, 185–186
AIDS (acquired immune deficiency
    syndrome)
  CCR5-Δ32 allele and resistance to,
    189–190, 203
  in human populations, 4
Åland Islands case study, 239–241
Albumin, in analyzing ape–human
    relationship, 91
Allele counting method, 29–30, 45–46, 184
Allele frequencies, 131–132, 168, 203, 257
  changes over in time in, 45
  chi-square statistic and, 47–48
  computing, 25–30
  deviations from random mating and, 36,
    38–39, 44
  Hardy–Weinberg equilibrium and,
    24–40, 45–48
  in impact of inbreeding on genotype
    frequencies, 63

Human Population Genetics, First Edition. John H. Relethford.
© 2012 Wiley-Blackwell. Published 2012 by John Wiley & Sons, Inc.